D0992425

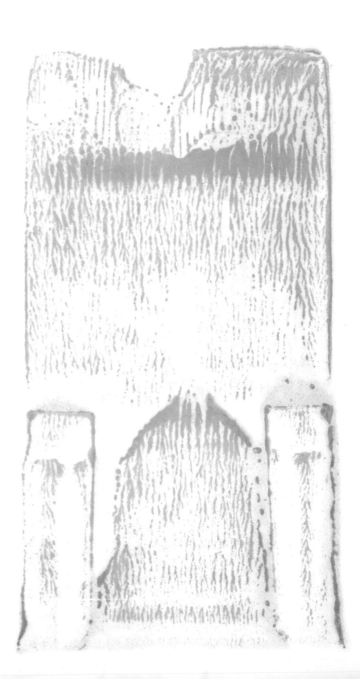

FLOWCHARTING

Mario V. Farina

Telecommunications & Information Processing Operations
General Electric Company

PRENTICE-HALL, INC., Englewood Cliffs, New Jersey

To
JOSEPH L. KATZ
who long ago taught me the importance
of flowcharting

Current printing (last digit):
10 9 8 7 6 5 4 3 2 1

(P) 13-322750-2

(C) 13-322768-5

Library of Congress Catalog Card Number: 79-112972
Printed in the United States of America

Prentice-Hall International, Inc., London
Prentice-Hall of Australia, Pty. Ltd., Sydney
Prentice-Hall of Canada, Ltd., Toronto
Prentice-Hall of India Private Ltd., New Delhi
Prentice-Hall of Japan, Inc., Tokyo

Preface

When you take a computer programming course, you will undoubtedly be advised by your instructor to "flowchart the solutions to your problems before you code them."

More often than not, however, the instructor will not say very much about flowcharting. Somehow, he either expects that the art of flowcharting is self-evident and doesn't require much explanation, or that it is something you can pick up on your own. It seems to me that most instructors don't want to be sidetracked from their main objectives - teaching you computing languages and techniques.

Flowcharting is *not* self evident. It is an art. One gets good at flowcharting the same way that he gets good at anything else - by practicing. However, some formal instruction is highly desirable. This text provides that initial instruction which will enable you to obey the instructor's dictum: *flowchart before coding.*

Part I of this text is fairly straightforward and you should be able to cover the material in about 2 hours. Try to complete this part as soon as possible before, or soon after, you begin your study of a programming language.

Part II is more complex. You should begin this part *after* you have begun studying *subscripts* in the programming language you are studying. The material in Part II should require about 12 hours of self-study, less if you have an instructor.

The BASIC programming language is used to help illustrate the flowcharting principles presented. I chose BASIC because it's easy to learn (it almost teaches itself) and effectively illustrates how actual programs can be easily written once flowcharts have been developed.

The text also serves as a brief introduction to programming techniques. At the beginning of the book the concepts and problems are very simple, but toward the end they become more complex. There should be much in those latter lessons which you can adapt for use in your own problem-solving activities.

Be sure to note the answers at the back of the book and the comprehensive index.

I am indebted to many persons who helped me prepare this text. Some of those persons are David N. Toussaint, who gave me some valuable advice; Joseph Tocco who carefully proofread the material; and Dante J. Pellei who offered suggestions for improvement. I'm especially grateful for the assistance given by Miss Nancy D. George who drew the flowcharts and provided the solutions to all the problems using General Electric's GE-265 System (Mark I); to Mr. Alvin J. Stehling, Chairman Data Processing, San Antonio College, who was the first to see and constructively comment upon the manuscript; and to Mrs. Beatrice Shaffer, who kept the IBM Selectric humming.

Mario V. Farina

Contents

PART I

PART II

BASIC in a Nutshell

(As Used in This Book)

To read value/s from a DATA line

  READ A
  READ P,Q,X

To accept data value/s from the keyboard

  INPUT E
  INPUT F,G,T

To assign a value to a variable

  LET X = 3
  LET D = A + B + 6

To print message/s or answer/s

  PRINT "END OF PROGRAM"
  PRINT X,L,K
  PRINT "VALUE IS",M,"AND",N

To branch unconditionally to another part of the program

  GO TO 385

To branch conditionally to another part of the program

IF A > B THEN 8	(greater than)
IF B < 3 THEN 28	(less than)
IF M = N * D THEN 305	(equal)
IF (P * Q)/L >= S THEN 99	(greater than or equa
IF 435 <= F THEN 30	(less than or equal)
IF J <> K THEN 25	(not equal)

To stop execution of a program

  STOP

To establish space for list/s

          DIM X(100)
          DIM K(50), V(25)

To set up a source of data value/s

          DATA -3,8,7,-6,0
          DATA 27

To use files (See Lesson 14 for details)

          FILES ALPHA;BETA
          READ #1,X
          SCRATCH #2
          WRITE #2,R;D;T
          IF END #1 THEN 86

Arithmetic Operators

+	Add	Examples:	LET A = B + 6
-	Subtract		LET D = E - F - G
*	Multiply		LET H = T * W * 3
/	Divide		LET P = (M/L)/7
↑	Raise to a power		LET Q = R↑2
			LET T = P↑K

Functions

SQR	Square root	Examples:	LET D = SQR(H * L)
SIN	Sine		LET N = SIN(F)
INT	Integer part		LET P = INT(K - W)
ABS	Absolute value		LET Q = ABS(175/B)

PART I

Lesson 1

BASIC CONCEPTS

A flowchart is a pictorial plan showing *what you want the computer to do and in what sequence.*

Let's start with a very simple example. Suppose you want the computer to compute 2 times 3, then print out the answer. Here's the flowchart:

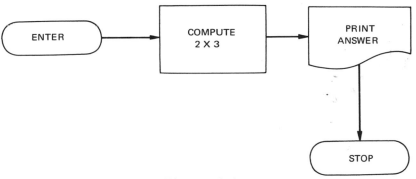

Figure 1-1

You can see that there are three different symbols shown in the flowchart. They are:

Each has a special purpose which we'll be discussing as we go along.

The purpose of a flowchart is to improve communications between one person and another. Through the technique of flowcharting he not only draws up a plan for himself to use for the instruction of a computer, but also communicates that plan to others.

So that confusion is held to a minimum, certain standards are observed by people who write flowcharts. This text follows the conventions outlined by US Standard Publication X3.5-1968.

Computer manufacturers provide templates that programmers may use. Figure 1-2 illustrates IBM Flowcharting Template X20-8020.

In Figure 1-1, the first symbol shown is:

It is called the *terminal* symbol. Its purpose is to show where a program begins and where it ends. Every flowchart should begin with this symbol and it should be located in the upper left-hand corner of the first page of your flowchart.

The symbol shown next in Figure 1-1 is

It is the *process* symbol. The symbol represents a processing function, such as a calculation.

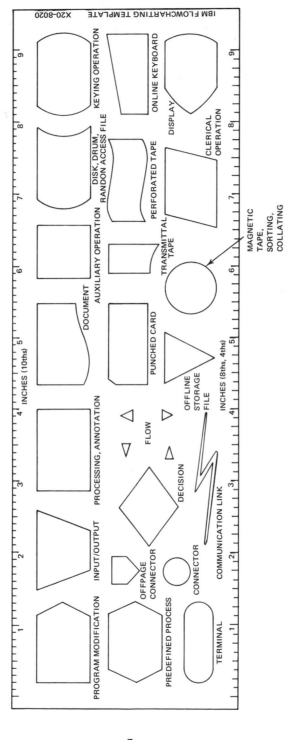

Figure 1-2

-3-

The third symbol in Figure 1-1 is

It is the *document* symbol. It represents a function
where the output from the computer is a document. In Figure 1-1, the document symbol shows that the answer to the
problem is printed on a sheet of paper.

The arrows connecting the shapes show *in what order*
the various functions have to be performed. As a general
rule, the flow of processing should be shown from left to
right and from top to bottom. It is not always convenient
to follow this rule; therefore, close attention should be
given to making sure arrows point in the right directions.

Do not let the simplicity of the example deceive you.
Flowcharts are virtually always more complex than the one
shown. A single flowchart often covers many pages.

In this text, we shall use the BASIC language to
illustrate the coding resulting from a flowchart. We'll do
this because BASIC is easy to understand even without a formal explanation. In your programming, you would, of course,
use whatever language you already know or one you are studying.

Here is the BASIC coding for the flowchart shown in
Figure 1-1.

```
10    LET C = 2 * 3
20    PRINT C
30    STOP
```

Note that the symbol marked ENTER has no corresponding BASIC instruction. The only use for ENTER is to show others where to begin reading your flowchart.

Make sure that you tell the computer what to do in the correct order. The flowchart shown in Figure 1-3 is wrong. Can you see why?

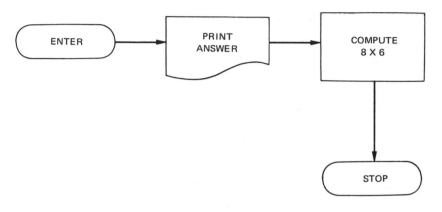

Figure 1-3

We're putting the cart before the horse. We're trying to print the answer before it is computed. This is, of course, absurd. You may smile, if you want to; however, programmers inadvertently do things like this more often than they like to admit.

Here's an important point: Knowledge of how to write flowcharts does not imply knowledge of *how to program*. When given a problem, *you* are the one who must come up with the procedure that the computer must follow in order to solve the problem. You should study the method used because it will give you clues about how to proceed in the future should you be given a similar problem.

Exercises

1. You want the computer to compute 5 x 8, then print out the answer. Write a flowchart to show this.

2. What's wrong with the flowchart in Figure 1-4?

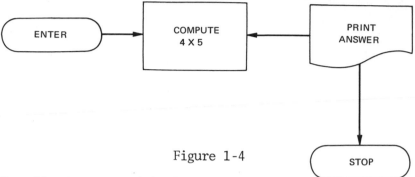

Figure 1-4

3. What's wrong with this flowchart?

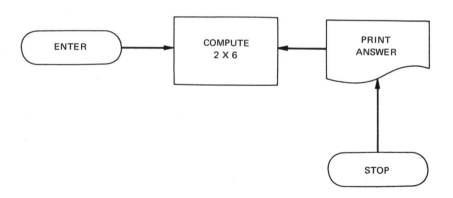

Figure 1-5

4. Rewrite the flowchart in exercise 3 so that it is correct.

Lesson 2

LOOPS

In the first lesson, the flowcharts shown were very simple. They showed several kinds of symbols connected with arrows. In the examples, once the function indicated in any one symbol has been performed, that function was never repeated.

A flowchart can, however, curve back on itself. Figure 2-1 shows an example:

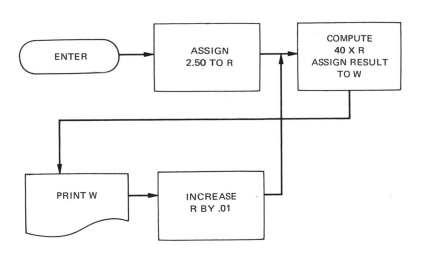

Figure 2-1

Here's the BASIC program that agrees with the flowchart in Figure 2-1.

```
10    LET R = 2.50
20    LET W = 40 * R
30    PRINT W
40    LET R = R + .01
50    GO TO 20
```

It would be wrong to write the last line as

```
50    GO TO 10
```

Can you see why? The program would not be written in accordance with the flowchart. The flowchart clearly shows that the program must go back to the symbol reading "COMPUTE 40 X R ASSIGN ANSWER TO W," *not* to the one reading "ASSIGN 2.50 TO R."

The example is simple but there is much to be learned. Note how the arrow flowing out of the bottom block in the flowchart curves back toward the beginning of the flowchart. This forms a "loop."

It's OK to show the returning arrow pointing directly to a block instead of ahead of it. The preceding flowchart could have been written as shown in Figure 2-2.

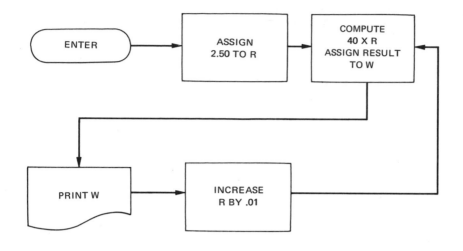

Figure 2-2

Be careful where you show a returning arrow. If you point it incorrectly your program will be coded incorrectly. Do your thinking as you develop your flowchart and your battle will be more than half won.

The phrases used in the blocks of a flowchart need not have any formal form. Whatever English words you care to use are OK so long as they clearly tell the computer what to do.

For instance, the phrases shown below all mean the same thing and could have been used in the first block of the flowchart.

```
ASSIGN 2.50 TO R
SET R EQUAL TO 2.50
INITIALIZE R WITH 2.50
PUT 2.50 IN R
R = 2.50
2.50 → R
```

You'll find that no two people will write a flowchart in exactly the same way. One programmer may be as brief and cryptic as this:

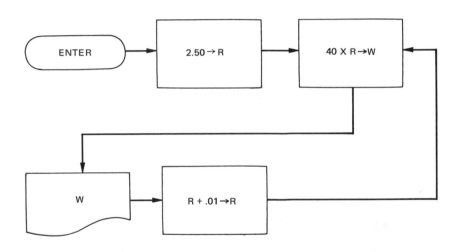

Figure 2-3

Observe that this programmer doesn't even use the word PRINT in the symbol showing W. He feels it is not necessary to write something that is *understood*. The *shape* of the symbol indicates clearly that the value of W is to be printed.

Another programmer may feel that a flowchart should make everything crystal clear. Therefore, his flowchart for the same problem may look like the one shown in Figure 2-4.

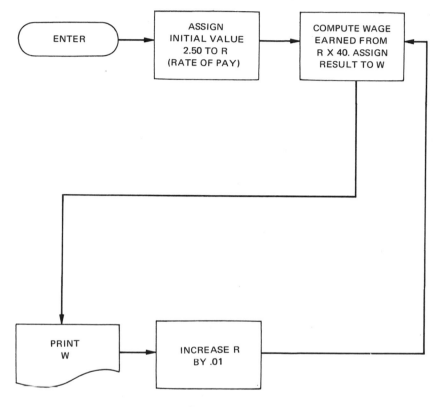

Figure 2-4

Sometimes a programmer will fill in the blocks with words and phrases closely related to the programming language he is using. For instance, this flowchart could be written as shown in Figure 2-5.

Notice how closely the phrases in the blocks agree with the program. You can see more clearly now why it is so truly said that when a flowchart has been developed, the program almost writes itself.

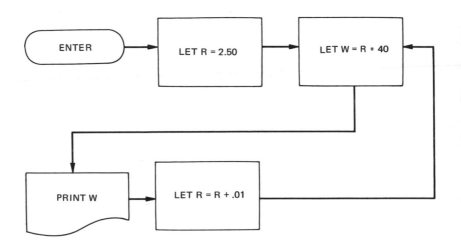

Figure 2-5

Note that the flowchart shows no block equivalent to GO TO 20. The *arrow* going from LET R = R + .01 to LET W = R * 40 takes the place of GO TO. In other words, a symbol reading GO TO is never necessary in a flowchart. *Arrows* show what the computer is to do and *in what sequence*.

Once a programmer has written a program in accordance with a flowchart, he may go back to the flowchart and write on it cross-references to the program. The flowchart last shown could be improved by writing on it the cross-reference shown in Figure 2-6.

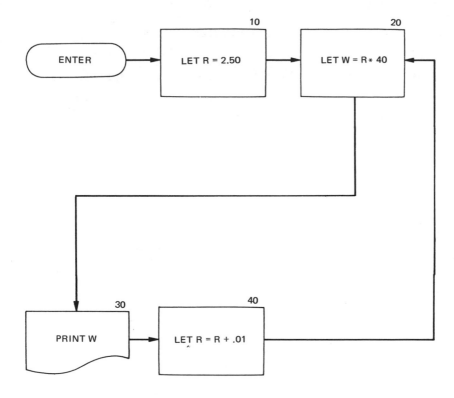

Figure 2-6

The numbers shown to the right above the symbols are called *line numbers* in BASIC. They would be *statement numbers* in FORTRAN, *procedure names* in COBOL, etc. Cross-referencing serves as documentation and helps in debugging if the program doesn't work.

You probably observed that all the examples in this lesson included no symbols reading STOP. The reason is that the flowcharts showed no way to terminate executions. Programs that do not show definite ways to stop are not good programs. Once begun, they run until they are stopped manually. The next lesson will show you how a program can test for its own termination.

Exercises

1. Write a flowchart for a program that is to compute salesmen's commissions of 15% for merchandise varying in price from $2.00 upward in steps of $.10. The program should include a loop with no definite termination condition.

2. Write the flowchart in Figure 2-7 in a different way. That is, use different phrases within the symbols.

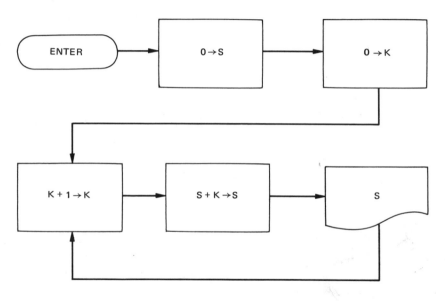

Figure 2-7

3. What's wrong with this flowchart?

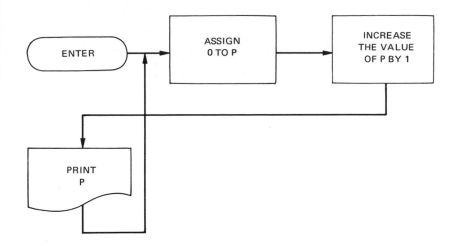

Figure 2-8

4. Draw a flowchart corresponding with this BASIC
   program

        10   LET L = 1000

        20   LET I = .075 * L

        30   PRINT I

        40   LET L = L + 100

        50   GO TO 20

In this program, L means *amount of loan*, I means
interest cost.

Lesson 3

MAKING DECISIONS

You can show alternate courses of action by writing
a symbol with a diamond shape

The test to be made is shown inside the diamond. Two or
more arrows flow from the symbol.

Suppose a problem is to compute tax at 20% for all
yearly earnings in the range of $4000 thru $9000 in steps
of $500. Figure 3-1 shows how the computer would be in-
structed to solve the problem.

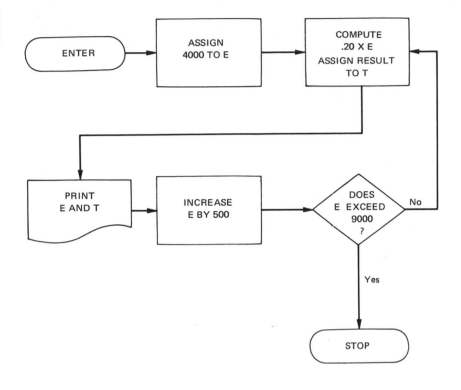

Figure 3-1

Observe the labels Yes and No that are appended to the arrows flowing from the diamond-shaped symbol. These arrows tell what the computer is to do when the question asked is answered. A few moments' thought will reveal that the question within the diamond will be asked 11 times. The answer will be No ten times, Yes once.

Here's the BASIC program corresponding to the flow-chart.

```
10    LET E = 4000
20    LET T = .20 * E
30    PRINT E, T
40    LET E = E + 500
50    IF E > 9000 THEN 70
60    GO TO 20
70    STOP
```

The statement at line 30 shows that you can direct the computer to print out more than one value at a time.

The statement at line 50 shows how a test is made. If the result of the test is *true* (E *is* greater than 9000), the job is done and the computer is directed to STOP. Otherwise, the computer will continue by calculating another tax value.

In the BASIC language, these symbols can be used

>   means greater than

<   means less than

>=   means greater than or equals

<=   means less than or equals

=   means equals

<>   means not equal

Let's take another example from a business situation. Suppose we have a milling machine that originally cost $30,000. We know it depreciates at a rate of 10% per year based upon its last reported value. That is, the first year its depreciation is 10% of $30,000 or 3000; the second year its depreciation is 10% of $27,000 ($30,000 - $3000); etc.

The problem is to make a table showing year and machine value until the value of the machine falls below scrap value, $2000. Figure 3-2 shows a flowchart we can use.

-18-

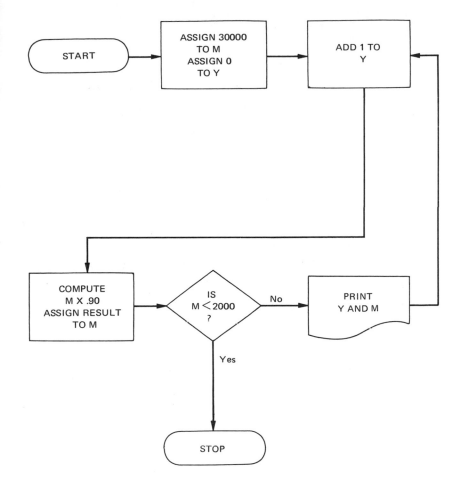

Figure 3-2

The letter M means *machine's value* and Y means *year*. Observe where Y is increased by 1. Here's the BASIC coding for this problem:

         10   LET M = 30000
         20   LET Y = 0
         30   LET Y = Y + 1

(Continued on next page)

-19-

```
40    LET M = M * .90
50    IF M < 2000 THEN 80
60    PRINT Y, M
70    GO TO 30
80    STOP
```

You may already have noticed that as you work in the development of a flowchart, your thoughts about how a problem is to be solved become clearer. You see things that were not obvious when you first began working at the solution. You may find that some initial thoughts you had were completely wrong, and you therefore have to begin the flowchart all over again. Do not let this bother you; it's perfectly normal. When you work on a flowchart, you need lots of paper, a sharp pencil, and a good eraser.

In BASIC, the DATA statement permits one to *read* values and assign them to variables. Here's a simple program illustrating how this is done:

```
10    READ S
20    IF S = 0 THEN 50
30    PRINT S
40    GO TO 10
50    STOP
60    DATA 3, 18, 24, -4, 0
```

The computer will print four lines of output: 3, 18, 24, and -4.

Suppose today we have sold three appliances: a refrigerator for $550, a TV set for $375, and an air-conditioner for $250. We are entitled to a commission of 15% on the first item, 12% on the second item, and 7.5% on the third item. How can we have the computer tell us what our individual commissions are? Here's a flowchart we can use to solve the problem.

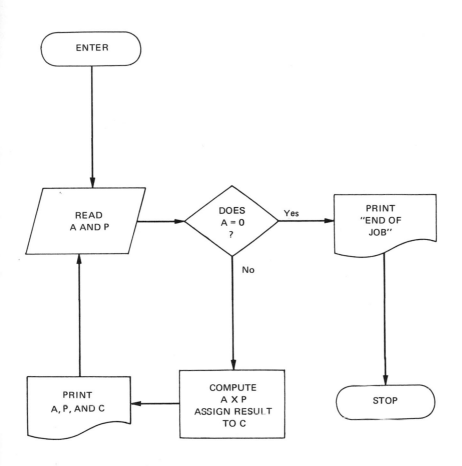

Figure 3-3

A new symbol appears in this example

It is the *input/output* symbol and represents information being read into the program for processing or information being written out. This symbol may be used for showing printed output in place of the document symbol, if desired.

The preceding flowchart shows the data names A, P, and C. The letter A means *appliance cost*, P means *percentage for commission*, and C means *commission*. The BASIC rendition of this program is:

```
10      READ A, P
20      IF A = 0 THEN 50
30      LET C = A * P
35      PRINT A,P,C
40      GO TO 10
50      PRINT "END OF JOB"
60      STOP
70      DATA 500, .15, 375, .12, 250, .075, 0, 0
```

Line 10 shows that in BASIC, sets of two or more values may be read at one time from the DATA statement. The DATA statement shows a set of dummy values "0" and "0" to terminate the execution of the program. (A *complete* set of values is needed despite the fact that the program tests only A.) At line 50, observe how, in BASIC, messages are printed.

Exercises

1. A retail store selling books and sundry items allows a 10% courtesy discount to teachers. Write a flow-chart that shows how the computer is to calculate and print out 10% of $1.00, $1.10, $1.20, etc., through $25.00.

2. Assume several typewriter part numbers are listed in a DATA statement. (Example: DATA 3014, 3029, 3083, 4017, 4044, 9999.) Write a flowchart showing how a computer program would test the numbers to see if they are in increasing sequence. The program is to print out NUMBERS IN SEQUENCE if all numbers are in sequence; otherwise, the program is to stop if even one number is out of sequence.

3. Assume you have a set of examination scores. Write a flowchart and a BASIC program to compute the average of the scores. The value, 999, should be included as a dummy score and used only to check for the end of the data items. It must not be included in the calculation.

4. This problem is to read a set of pay numbers, X, Y, and Z, from the DATA statement and have the computer determine which of the three is the largest. When that task has been accomplished, it should repeat the process with three more pay numbers, etc. The program is to repeat this procedure until X's value is found to be 999. Do not process this case; simply stop the program. Write the flowchart and BASIC program that solves this problem. Make sure that the last set of values is 999, 0, 0. In BASIC, you cannot read a partial set of values.

Lesson 4

NEATNESS COUNTS

A flowchart is a work of art. No two flowcharts will
show the solution to a problem in identically the same way.
When you flowchart, do not hesitate to be creative; however,
do try to use the standards described in this text and do
try to plan your flowcharts so that the flow of processing
is shown proceeding from left to right and from top to bottom. Doing this will enable others to understand better the
procedures you develop for the solution of problems. In this
lesson we'll discuss some of the techniques you can employ
that will help stamp your flowcharts as the work of a pro-
fessional.

It's all right to show more than one task being performed in a processing symbol. See Figure 4-1.

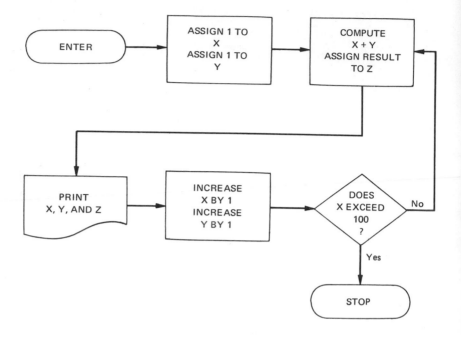

Figure 4-1

The first block following the START symbol shows initial values assigned to X and Y. Both of these functions must be completed before the calculation in the next block is performed. The functions must be coded in the same order in which they appear in the symbol. You'll notice that two functions are also performed in the processing symbol, which reads INCREASE X BY 1, INCREASE Y BY 1.

Sometimes a programmer will run out of space on a sheet he's working on and may have to improvise a bit. For example, the preceding flowchart could be written this way:

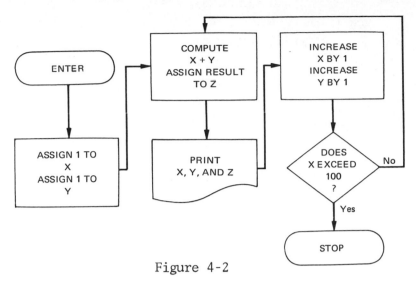

Figure 4-2

In either case, the BASIC coding for this program will be as follows:

```
10   LET X = 1
20   LET Y = 1
30   LET Z = X + Y
40   PRINT X, Y, Z
50   LET X = X + 1
60   LET Y = Y + 1
70   IF X > 100 THEN 90
80   GO TO 30
90   STOP
```

You can see that this program computes 1 + 1, 2 + 2, 3 + 3,..., 100 + 100. The program shows nine statements written *serially* down the length of the paper, but the flowchart could be squeezed in a limited space. Nevertheless, the flowchart corresponds perfectly with the program. The arrows on the flowchart leave no doubt about the order in which statements of the program are to be executed.

-27-

Even if the flowchart had been written in the confusing form shown in Figure 4-3, BASIC coding would remain the same.

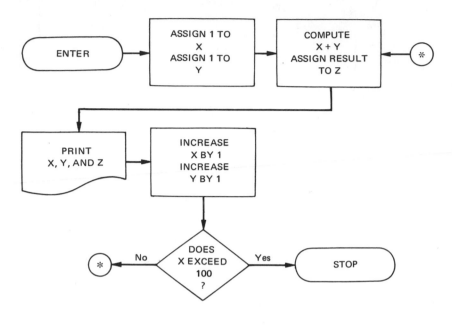

Figure 4-3

The arrows show what function must be performed and in what order. We're certainly not suggesting that you write flowcharts as messy as this one. Sometimes, however, to conserve space, to keep related blocks near each other, or to avoid crossing lines, you may have to resort to expediencies such as these.

The flowchart shows a new symbol

It is the *connector* symbol, which is used to connect one part of a flowchart with another. Sometimes those parts are on different pages.

In Figure 4-3, we see

To see where the flowchart continues from this point we must
find another circle with an enclosed asterisk. We find it
in the upper right-hand corner.

Connectors can often be avoided, of course, by drawing
arrows from one part of the flowchart to a distant part.
Doing so, however, tends to make a complicated flowchart appear cluttered.

You can enclose *any symbol* or alphabetic character you
please within connector symbols. Connectors such as

are OK.

If a connector and its corresponding connector are on a
different page of a flowchart, the connectors should show
page references. For instance, connectors might appear like
this:

Exercises

1. Rewrite this flowchart to make it neater.

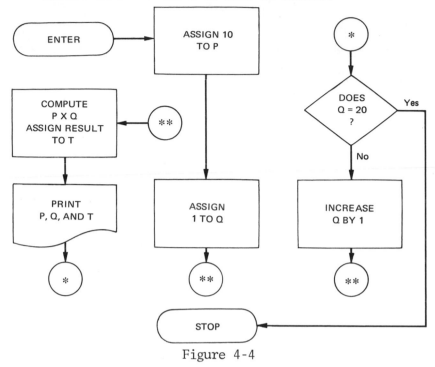

Figure 4-4

2. Write the BASIC program to agree with your flow-chart for exercise 1.

3. Review the names of the following symbols and their uses.

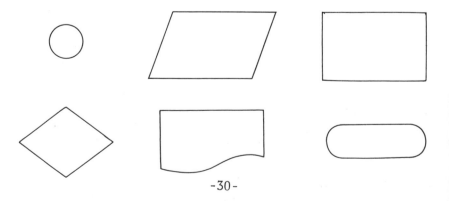

Lesson 5

READING DATA

The ability of a computer to make decisions and to loop is very important. Without these abilities, it is unlikely that computers would be used to do even 1% of the work they now do.

Another powerful feature of computers is the ability to *read* data. Reading data means that information in auxiliary devices, such as card readers, tape handlers, and disc files, is transferred to the computer's memory at the request of programs. Recall that we've already discussed how data are read in BASIC.

Assume there is a deck of punched cards in a computer's card reader. Each card holds three numbers. Let's write the flowchart for a program to read the information on those cards, one card at a time, and to print out that information.

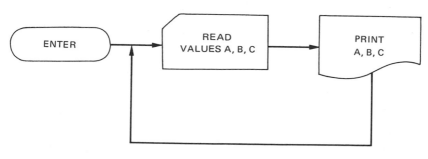

Figure 5-1

The flowchart introduces a new symbol, the *punched card* symbol.

Whenever you want to indicate an instruction for reading cards, you can use the standard input/output symbol. Do you remember the programmer who didn't like to do much writing in flowcharts? He let the *shapes* of the symbols speak for themselves. He might, therefore, write the above flowchart this way:

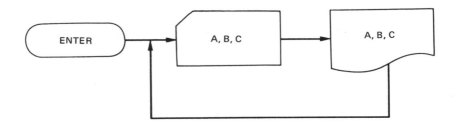

Figure 5-2

This flowchart shows no definite termination condition for the program.

Let's assume that the *last* data card is a dummy card (not to be processed) and holds the value 999 for A. Now we can have the flowchart show a test for this card. Once the computer has found this card, it can jump to a statement that either stops the program or does further processing. Here's an example where the computer is simply told to print END OF JOB when the dummy card has been found.

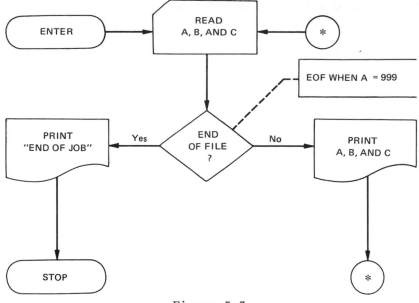

Figure 5-3

A FORTRAN program that could be written in accordance with this program is:

```
35   READ 15, A, B, C
15   FORMAT (3F10.3)
     IF (A.EQ.999.) GO TO 50
     PRINT 15, A, B, C
     GO TO 35
50   PRINT 55
55   FORMAT (11H END OF JOB)
     STOP
     END
```

You may wonder why we've suddenly switched to FORTRAN where previously we used BASIC to illustrate the coding resulting from flowcharts. The reason is that BASIC, being a teletype-oriented language, does not read punched cards. Instead, data are built right into programs. On the other

-33-

hand, reading cards is a normal mode of operation in FORTRAN.
Here is the same program in BASIC.

```
10    READ A, B, C
20    IF A = 999 THEN 50
30    PRINT A, B, C
40    GO TO 10
50    PRINT "END OF JOB"
60    STOP
70    DATA 6, 7, 4, 2, 9, 17, 4, 6, 3, 999, 0, 0
```

Although this problem is simple, there are several new points
to be noted. First, the decision diamond does not show what
is meant by "end-of-file." It is not necessary for the diamond to show this if that information appears elsewhere on
the chart. Observe the note to the right of the chart which
defines the EOF (end-of-file) condition. You'll see later
that flowcharts make extensive use of notes to give explana-
tory material that will not fit in flowchart blocks. For instance, complicated equations are often shown in notes.

Notes are enclosed in the symbol provided for that purpose, the *annotation* symbol.

It is connected to any flowchart symbol at a point where the
annotation is meaningful.

In line 70, you'll see a series of values. These are
assigned to A, B, and C as the program calls for them. (There
are three actual sets of values used. The last set (999, 0, 0)
is not processed.) When the program sees that A equals 999,
it jumps to line 50, prints "END OF JOB" and stops. The
two trailing zeros of the dummy set are necessary, as we've

noted before, because a BASIC READ statement cannot read a partial set of values. From this point, we will use the standard input/output symbol to indicate reading data.

Let's solve another problem. Assume there are several values to be read from a data source. The values are to be added and the sum printed. Figure 5-4 is a flowchart that shows the procedure to accomplish the task. Assume that the end-of-file (dummy) value is zero.

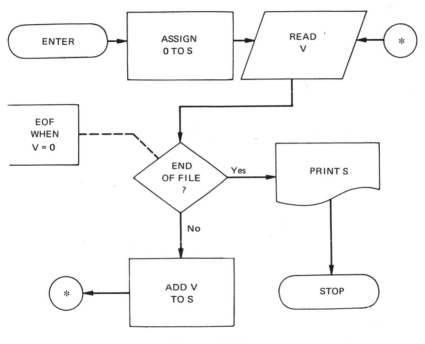

Figure 5-4

Note that the processing symbol shows zero assigned to S. We want to be sure S starts off with a "clean slate." Therefore, we have the computer *initialize* S. The initialization portion of a program is located at or near the beginning of the program. Observe that initialization symbols are outside the main body of loops.

-35-

Here's the BASIC program that corresponds with the preceding flowchart:

```
10    LET S = 0
20    READ V
30    IF V = 0 THEN 60
40    LET S = S + V
50    GO TO 20
60    PRINT S
70    STOP
80    DATA 8, 3, 17, 4, 7, 20, 0
```

You can see that in the DATA statement, zero is the
last value. It acts as an end-of-file indicator and is not
processed. The person who selects a value to act as an end-
of-file indicator must make sure that this specific value
will *never* appear as one of the data items to be actually
processed.

Exercises

1. Write the flowchart corresponding with this program (this program finds the largest value in the
   DATA statement and prints it out).

   ```
   10    LET M = 0
   20    READ A
   30    IF A = 0 THEN 70
   40    IF M > A THEN 20
   50    LET M = A
   60    GO TO 20
   70    PRINT "LARGEST VALUE IS ", M
   80    STOP
   90    DATA 2, 8.9, 7.3, 1.7, 11.3, 4.2, 0
   ```

2. Assume end-of-file value, V, is 99. Do the
   decision diamonds in Figure 5-5 all ask the same
   question?

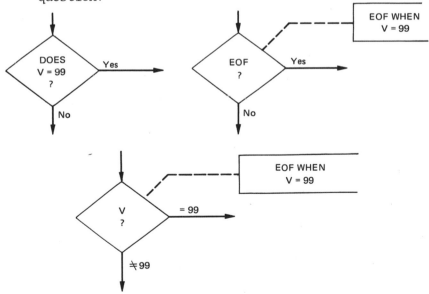

Figure 5-5

Lesson 6

BASIC DATA PROCESSING FLOWCHART

The heart of 90% of all flowcharts used for the solution of problems, whether scientific or business, has this basic pattern:

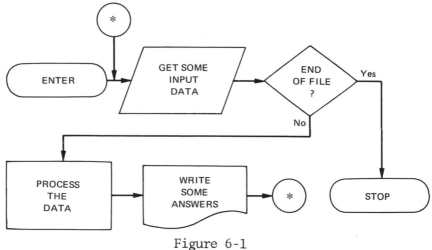

Figure 6-1

Given a job, you should try to find if the pattern exists, and, if so, to fill in the missing pieces.

Suppose a problem is to compute weekly gross salaries when hours worked and rates per hour are shown as data. Assume the end-of-file value is hours worked = 1000. Figure 6-2 shows a flowchart that can be used.

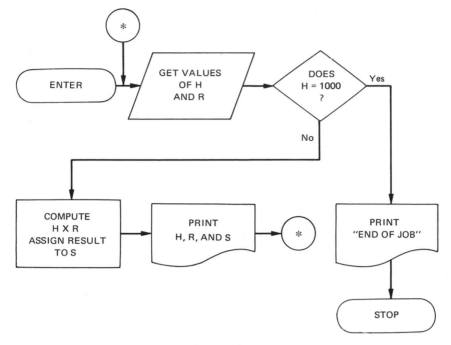

Figure 6-2

The flowchart follows the basic data processing pattern. You can probably detect the block that calls for the input data, the one that processes the data, and the one that writes answers.

Note that when the end-of-file is detected, it is not always required that the program stop immediately. Some processing (often a great deal of processing) can be done after the end-of-file is found. A BASIC program that agrees with the preceding flowchart is as follows:

```
10    READ H, R
20    IF H = 1000 THEN 60
30    LET S = H * R
40    PRINT H, R, S
50    GO TO 10
```
(Continued on next page)

60   PRINT "END OF JOB"

70   STOP

80   DATA 40, 2.55, 36, 3.50, 25, 2.90, 1000, 0

As another illustration, assume a salesman wants to compute his total commission on the sale of various appliances. He gets a commission of 5% on items selling for less than $100 and 7.5% on items selling for $100 or more. The values of the sales are shown in the DATA source. Here's a flowchart illustrating one way to solve the problem:

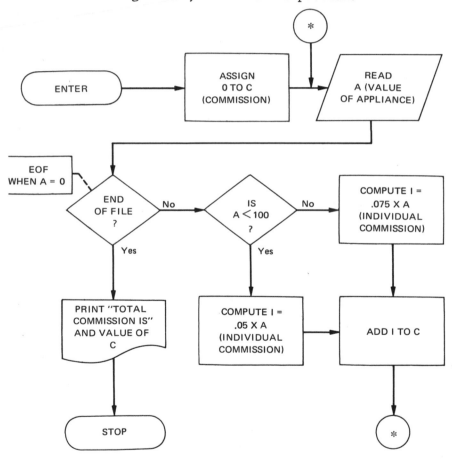

Figure 6-3

A BASIC program that agrees with this flowchart is this:

```
10   LET C = 0
20   READ A
30   IF A = 0 THEN 100
40   IF A < 100 THEN 70
50   LET I = .075 * A
60   GO TO 80
70   LET I = .05 * A
80   LET C = C + I
90   GO TO 20
100  PRINT "TOTAL COMMISSION IS", C
110  STOP
120  DATA 45, 416, 93, 21, 157, 99, 101, 100, 0
```

(You can see that the end-of-file indicator is A = 0)

Exercises

1. Write the flowchart that agrees with this next
    BASIC program:

    ```
    10    PRINT "THIS PROGRAM COMPUTES SUM OF DIGITS"
    20    LET S = 0
    30    READ X
    40    IF X > 100 THEN 70
    50    LET S = S + X
    60    GO TO 30
    70    PRINT "ANSWER IS", S
    80    PRINT "END OF JOB"
    90    STOP
    100   DATA 8, 9, 44, 17, 21, 6, 44, 999
    ```

2. Identify the following in the preceding program:

    a.   the part where input data are read;
    b.   the part where data are processed;
    c.   the part where answers are written out;
    d.   the part that tests for end-of-file;
    e.   initializing statements.

3. A device being made in the shop is composed of
    several parts. The costs of these parts are
    listed in a data source. It is desired to sum
    the individual costs above $25. Write a flow-
    chart to show how this task is accomplished. The
    end-of-file dummy cost is 0.

4. Write a BASIC program agreeing with the flowchart
    in exercise 3.

Lesson 7

COUNTING

There will be times when you'll need to write a program where *counting* is to be performed. The following flowchart is for a program that does nothing but go through a do-nothing loop 5000 times.

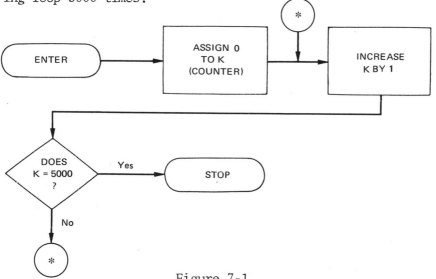

Figure 7-1

Despite the fact that this flowchart is simple, it is instructive. Note K. It is initialized with the value zero; then it is increased by 1. The instruction that increases K by 1 is in the body of a loop which is executed many times - in the example, 5000 times.

Now we can write a more meaningful program, using the preceding flowchart as a guide.

Suppose we want to calculate the sum of integers from
1 to 1000. There's an equation that gives this answer, but
let's forget it for now. Let's show a loop where the program itself counts the number of times it has executed the
loop (see Figure 7-2).

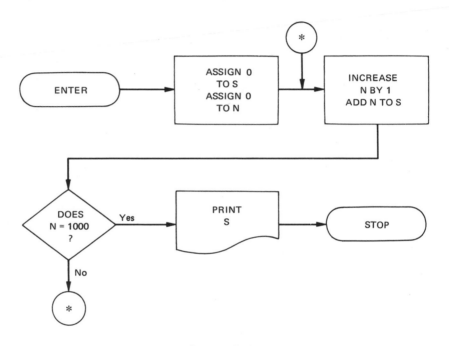

Figure 7-2

You can see that values of N will be 0 (its original
value), 1, 2, ..., 1000, and that values of S will be 0 (its
original value), 1, 3, 6, 10,..., 500500.

A BASIC program that agrees with the flowchart in Figure 7-2 is shown on the next page.

```
10    LET S = 0
20    LET N = 0
30    LET N = N + 1
40    LET S = S + N
50    IF N = 1000 THEN 70
60    GO TO 30
70    PRINT S
80    STOP
```

Some programmers prefer to place the test for the end of a loop near the beginning of the loop (see Figure 7-3).

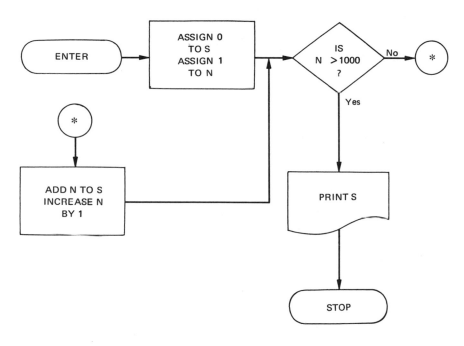

Figure 7-3

Others prefer to count the number of iterations this way:

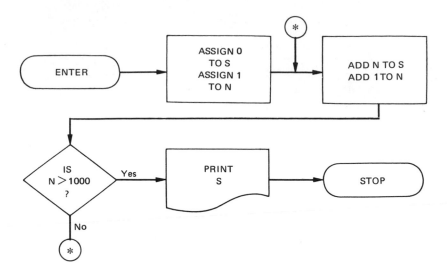

Figure 7-4

Can you see why the decision diamonds in the last two flowcharts ask whether N is *greater than* 1000? Still another way is shown in Figure 7-5.

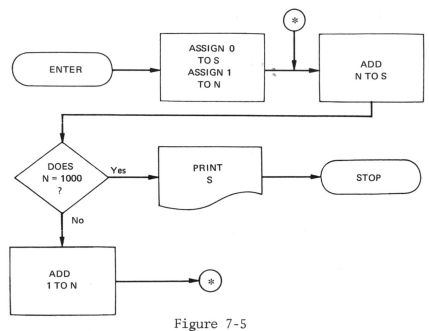

Figure 7-5

We didn't show you all these ways to count just to con-
fuse you but to show that there may sometimes be several ways
to accomplish the same task. When you solve problems, you
should use whatever method seems most natural to you. Sometimes you will have to reinitialize a counter. Let's say
there are 12 production costs in a data source. You want to
add them up in sets of three and print the sum of each set
(four answers). Figure 7-6 shows how you would do it.

You will notice that there is a *loop within a loop*. The
name N counts how many sets of three values have been processed; K counts the number of values in each set.

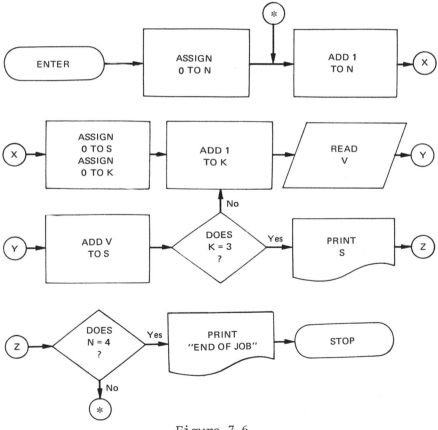

Figure 7-6

-47-

The following BASIC program corresponds with the flow-chart in Figure 7-6.

```
10    LET N = O
20    LET N = N + 1
30    LET S = 0
40    LET K = 0
50    LET K = K + 1
60    READ V
70    LET S = S + V
80    IF K = 3 THEN 100
90    GO TO 50
100   PRINT S
110   IF N = 4 THEN 130
120   GO TO 20
130   PRINT "END OF JOB"
140   STOP
150   DATA 3.25, 6.16, 4.94, 7.82, 8.01, 3.67
160   DATA 2.98, 9.06, 4.00, 6.21, 6.91, 8.22
```

Exercises

1. Write the flowchart that agrees with this BASIC program.

```
10    READ N
20    LET S = 0
30    LET K = 0
40    LET K = K + 1
50    READ V
60    LET S = S + V
70    IF K = N THEN 90
80    GO TO 40
```

(Continued on next page)

```
90    PRINT S
100   STOP
110   DATA 10, 3, 6, 7, 4, 8, 2, 1, 6, 2, 4
```

2. Assume that a data list contains 10 values in five
   sets of two. The problem is to multiply the two
   numbers in each set, then add the five products.
   The computer is to write out each individual
   product and also the sum of the products. Write
   a flowchart to show how this job would be done.

3. Write a flowchart to show the solution to this
   problem. Obtain the sum of odd integers from
   1 to 99, inclusive.

4. Write a BASIC program to agree with exercise 3.

5. A data source has several ages recorded. (The end-
   of-file dummy age is 0.) Write a flowchart to
   show how a program should count the number of
   ages below 12 and the number of ages that are 12
   or above.

6. A salesman has made several sales. He wants to
   know how many sales were $100 or less, how many
   were above $100 but not more than $200, and how
   many were above $200. Write a flowchart to show
   how a program should be written to give this information.

Lesson 8

TWO INSTRUCTIVE PROBLEMS

In this lesson, we wind up Part I of this text with two interesting and educational problems. These problems are for math and science students. Other students may skip them.

The first problem is to compute the roots of the quadratic equation

$$R = \frac{-b \pm \sqrt{b^2 - 4ac}}{2a}$$

given the values of a, b, and c. Assume the values a, b, and c are available in sets from a data source. Have the computer test for negative *discriminants*, $b^2 - 4ac$. If a discriminant is negative, have the computer disregard the case and go on to the next one; if the discriminant is zero or greater than zero, have the computer calculate roots R1 and R2.

Be sure also to have the computer test to see whether a is zero. If it is, division is impossible and the computer is to disregard the case. The end-of-file value a is 999.

In BASIC, one obtains the square root of a value by calling for the built-in function SQR and by placing an argument within parentheses. Here's an example.

LET W = SQR((P + Q)/R)

The computer will give the square root of $(P + Q)/R$.

Before looking at the flowchart in Figure 8-1, see if you can develop one of your own.

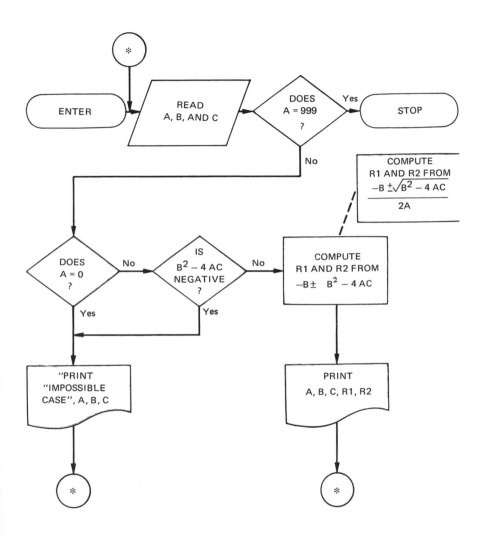

Figure 8-1

The program is easily written:

```
10    READ A, B, C
20    IF A = 999 THEN 90
30    IF A = 0 THEN 100
40    IF B ↑ 2 - 4 * A * C < 0 THEN 100
50    LET R1 = (-B + SQR(B ↑ 2 - 4 * A * C))/(2 * A)
60    LET R2 = (-B - SQR(B ↑ 2 - 4 * A * C))/(2 * A)
70    PRINT A, B, C, R1, R2
80    GO TO 10
90    STOP
100   PRINT "IMPOSSIBLE CASE", A, B, C
110   GO TO 10
120   DATA 6, 7, 4, 3, 9, 2, 8, 1, 7, 999, 0, 0
```

Mathematicians will not agree that impossible situations arise when discriminants are negative. They may, of course, modify the program to compute complex roots.

The second problem is to find the area under the sine curve from X = 0 to X = Π/2.

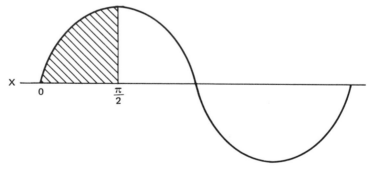

Figure 8-2

A calculus procedure can be used, but for now let's take a different approach. Divide the curve into horizontal strips.

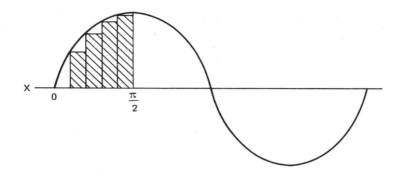

Figure 8-3

Then find the areas for each of the strips and compute the sum of the areas. This will be an *approximation* to the area required. The more strips you put under the curve, the more accurate will be the approximation. In the solution to this problem, make your strips Π/100 wide.

For this problem you'll need to know how to compute sines. Use the built-in function SIN; for example

LET D = SIN (M/E)

You can use this flowchart to solve the problem.

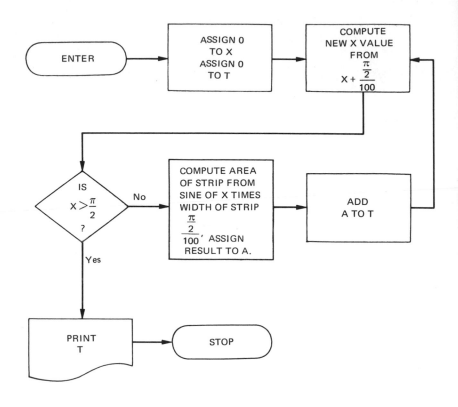

Figure 8-4

The program now almost writes itself.

```
10      LET X = 0
20      LET T = 0
30      LET X = X + (3.1416/2)/100
40      IF X > 3.1416/2 THEN 80
50      LET A = SIN(X) * ((3.1416/2)/100)
60      LET T = T + A
70      GO TO 30
80      PRINT T
90      STOP
```

-54-

1. Can you improve the preceding program so that it runs faster? Keep the width of the strips at Π/100.

2. Can you improve the preceding program so that it has fewer statements?

3. Can you improve the preceding program so that the area of each strip under the curve is computed at its mid-point, as shown?

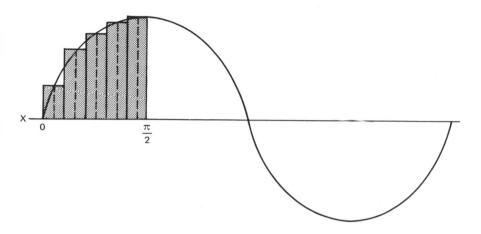

4. Write a flowchart that shows how a program would be written to compute all possible resultant resistance values from R1 and R2 connected in parallel. Note that R1 varies from 10 to 20 inclusive in steps of .5, and that R2 varies from 50 to 80 inclusive in steps of 1. The equation to compute parallel resistance is:

$$R = \frac{R_1 \times R_2}{R_1 + R_2}$$

5. Write a BASIC program agreeing with exercise 4.

PART II

Lesson 9

SUBSCRIPTS

Suppose you have a list of 20 values on a data source and you want the values stored in a list. In this lesson, problems using lists are going to get much more complex than that, but let's assume that that's all there is to the first problem. Figure 9-1 shows the way you'd write the flowchart.

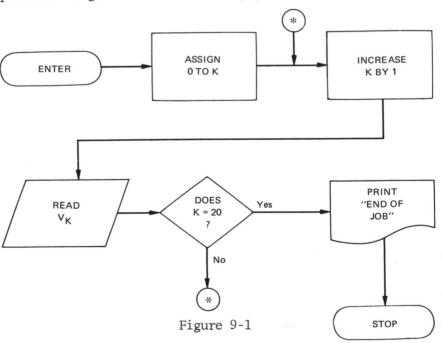

Figure 9-1

The name V does not represent just one value but an *array* of values - in this case, 20. The letter K is a subscript that *points* to the particular V value being accessed. When K is 1, the value being accessed is V_1; when K is 2, the value is V_2; etc.

```
10    DIM V(20)
20    LET K = 0
30    LET K = K + 1
40    READ V(K)
50    IF K = 20 THEN 70
60    GO TO 30
70    PRINT "END OF JOB"
80    STOP
90    DATA 6, 3, 9, 4, 7, 6, 3, 1, 4, 9, 8, 6
100   DATA 1, 6, 4, 7, 8, 5, 2, 4
```

Bear in mind that this illustrative program is meaningless; it doesn't show what happens to the list of 20 values once it has been set up.

However, this flowchart and program does show how subscripts work. In the example, K is a subscript. It varies from 1 through 20, and therefore, references V_1 through V_{20}. Note that in the BASIC language, as in most other major programming languages, subscripts are shown within parentheses.

Observe also the DIM statement in line 10 which merely *reserves* 20 computer memory positions for the array V. Most other major programming languages include statements similar to BASIC's DIM. For example, FORTRAN uses DIMENSION; COBOL uses OCCURS; etc. It is not necessary to show on a flowchart the reservation of space for an array.

Let's say our next problem is to read 20 age values into an array (call it C), then have the computer determine

-57-

what is the largest value in the array. Figure 9-2 shows a
flowchart we can use.

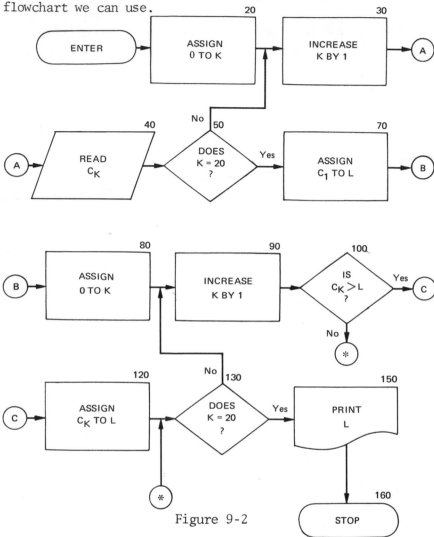

Figure 9-2

To solve the problem, we arbitrarily assume the first C
holds the largest age; then we test the remaining ages (the
second through the twentieth) to determine whether any of
them are larger. It's possible, therefore, for the age value held in L to be changed 19 times from its original setting.
Here's the program.

```
10    DIM C(20)
20    LET K = 0
30    LET K = K + 1          THIS PART OF THE PROGRAM
40    READ C(K)              READS IN 20 AGE VALUES
50    IF K = 20 THEN 70
60    GO TO 30
70    LET L = C(1)
80    LET K = 1
90    LET K = K + 1
100   IF C(K)>L THEN 120
110   GO TO 130              THIS PART OF THE PROGRAM
120   LET L = C(K)           FINDS THE LARGEST AGE
130   IF K = 20 THEN 150     AND PRINTS IT OUT
140   GO TO 90
150   PRINT L
160   STOP
170   DATA 3, 9, 4, 17, 8, 4, 12, 11, 17, 4, 3
180   DATA 7, 1, 8, 4, 9, 10, 9, 3, 2
```

This program should report that the largest age in the array is 17.

Suppose you need to know *which* of the 20 ages is the largest. We have to make only a slight modification to the flowchart shown earlier (See Figure 9-3).

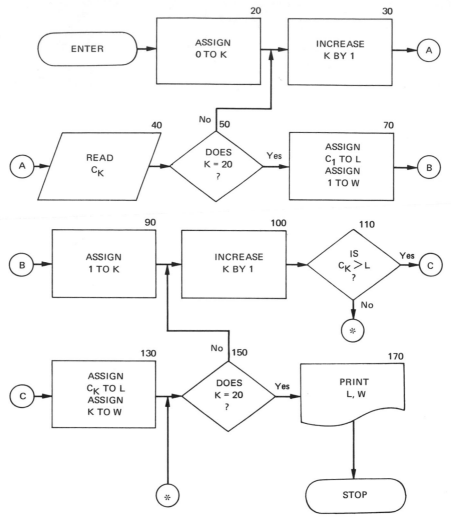

Figure 9-3

The letter W tells *which position* in the array contains the largest value. Note that its initial value is "1" (because C_1 is initially assigned to L). The value of W changes as the age value assigned to L changes.

This program should report that the largest age in the array is 17 and that it is the fourth value of the array.

Here's the BASIC program

```
 10   DIM C(20)
 20   LET K = 0
 30   LET K = K + 1
 40   READ C(K)
 50   IF K = 20 THEN 70
 60   GO TO 30
 70   LET L = C(1)
 80   LET W = 1
 90   LET K = 1
100   LET K = K + 1
110   IF C(K) > L THEN 130
120   GO TO 150
130   LET L = C(K)
140   LET W = K
150   IF K = 20 THEN 170
160   GO TO 100
170   PRINT L, W
180   STOP
190   DATA 3, 9, 4, 17, 8, 4, 12, 11, 17, 4, 3
200   DATA 7, 1, 8, 4, 9, 10, 9, 3, 2
```

Exercises

1. Have the computer read 20 catalog numbers into an
    array (call it C); then have the computer sort the
    catalogs from smallest to largest. Solve the prob-
    lem without setting up another array. Write the
    flowchart.

2. Write the BASIC program that corresponds to your
    flowchart in exercise 1.

Lesson 10

TABLE LOOK-UPS

A program may require that information be looked up in a table. Let's say that in our business we have an inventory of parts. We also have a computer that records in its memory (1) part numbers, (2) quantities on hand, and (3) selling prices. We want to write a program permitting us to interrogate the part records.

In some installations, the computer can be interrogated via teletype. For instance, in this example we can enter a part number on the keyboard and the program will look up and then type out the quantity on hand and the selling price for that part.

In real life, the inventory of parts could well contain thousands of items, but for our purposes, let's say it contains only 10.

You will notice a new shape in the flowchart in Figure 10-1.

This symbol is the *manual input* symbol which is used to represent a manual input function where the device employed is an online typewriter or teletype.

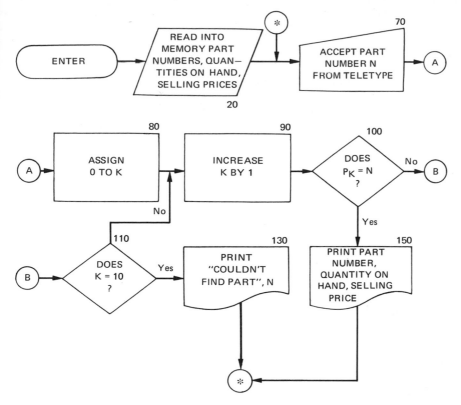

Figure 10-1

Here's a BASIC program written to agree with the preceding flowchart:

```
10  DIM P(10), Q(10), S(10)
20  LET K = 0
30  LET K = K + 1
40  READ P(K), Q(K), S(K)
50  IF K = 10 THEN 70
60  GO TO 30
70  INPUT N
80  LET K = 0
90  LET K = K + 1
```

(Continued on next page)

```
100   IF P(K) = N THEN 150
110   IF K = 10 THEN 130
120   GO TO 90
130   PRINT "COULDN'T FIND PART", N
140   GO TO 70
150   PRINT P(K), Q(K), S(K)
160   GO TO 70
170   DATA 500, 86, 1.25, 430, 160, 3.28, 610
180   DATA 3, 8.45, 247, 105, 2.45, 680, 11
190   DATA .75, 246, 92, 2.28, 199, 340, 7.05
200   DATA 360, 18, 1.65, 702, 84, 6.45, 225
210   DATA 75, 8.65
```

The INPUT statement in BASIC requires that the programmer type in one or more numbers required. In line 70, the computer requests the user to type in a part number. It will then wait until he does so.

You probably noticed that the flowchart showed one symbol that accounts for lines 20 thru 60 in the program. This is the general input/output symbol. It's all right to show this much activity in a single symbol if the task shown is routine and causes no confusion.

This program is stopped "manually." When the computer requests a part number, the user may type STOP. This automatically terminates the execution of the program.

Now let's try a problem where two lists are involved - a temperature list, T, and a pressure list, P. There are 15 entries in each list. These entries are found in the program's DATA statement in sets (each temperature value is matched with a corresponding pressure value). The temperature values are in random order.

Our first task will be to read in temperature-pressure sets, then sort the sets in increasing sequence according to temperatures. (The two lists vary together, so that when you've sorted temperatures in increasing sequence you've also sorted pressures in increasing sequence.) In the last lesson you learned how to sort; therefore, this first task will be relatively easy. Figure 10-2 shows the flowchart.

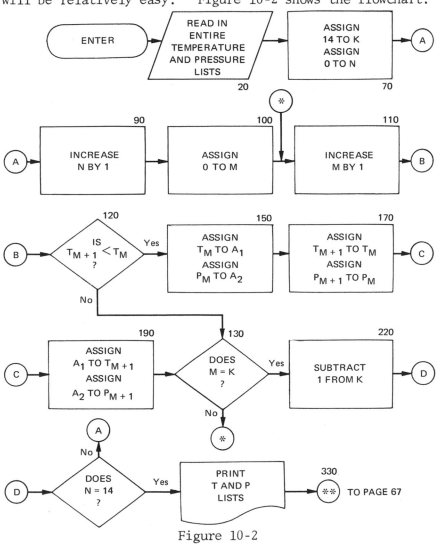

Figure 10-2

The BASIC program so far is this.

```
 10   DIM T(15), P(15)
 20   LET K = 0
 30   LET K = K + 1
 40   READ T(K), P(K)
 50   IF K = 15 THEN 70
 60   GO TO 30
 70   LET K = 14
 80   LET N = 0
 90   LET N = N + 1
100   LET M = 0
110   LET M = M + 1
120   IF T(M + 1) < T(M) THEN 150
130   IF M = K THEN 220
140   GO TO 110
150   LET A1 = T(M)
160   LET A2 = P(M)
170   LET T(M) = T(M + 1)
180   LET P(M) = P(M + 1)
190   LET T(M + 1) = A1
200   LET P(M + 1) = A2
210   GO TO 130
220   LET K = K - 1
230   IF N = 14 THEN 250
240   GO TO 90
250   LET K = 0
260   LET K = K + 1
270   PRINT T(K), P(K)
280   IF K = 15 THEN 330
290   GO TO 260
300   DATA 10, 20, 100, 158, 95, 120, 15, 30
```

(Continued on next page)

```
310   DATA 80, 105, 35, 62, 75, 100, 85, 106
320   DATA 40, 63, 50, 75, 105, 180, 110, 200
325   DATA 20, 40, 55, 80, 108, 190
```

Now we want to add the portion of the flowchart that tells
what to do *after* the sorting has taken place (see Figure
10-3).

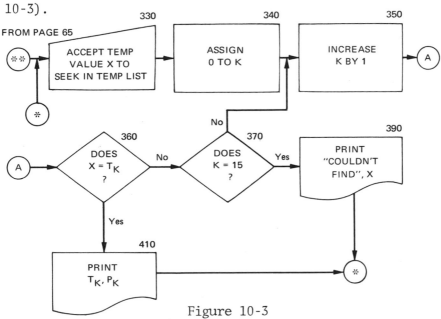

Figure 10-3

The BASIC coding begins with line 330.

```
330   INPUT X
340   LET K = 0
350   LET K = K + 1
360   IF X = T(K) THEN 410
370   IF K = 15 THEN 390
380   GO TO 350
390   PRINT "COULDN'T FIND", X
400   GO TO 330
410   PRINT T(K), P(K)
420   GO TO 330
```

The program is stopped by typing STOP whenever the computer asks for a temperature value.

Exercises

1. Assume that the ABC Cork Company has developed a table that shows how its net profit is expected to vary with respect to sales. Data are available in a data source in sets. For example, the sets begin this way.

    24, 124, 32.5, 130, 36, 136, 38, 142, 40, 148, ...

    where 24, 32.5, 36, etc., represent profit figures (in thousands of dollars), and 124, 130, 136, etc., represent sales figures (in thousands). Write a flowchart that shows how a program would give the profit for a given sales value input from the keyboard of a teletype. The flowchart would have to show how a sales value not in the sales list (such as 139) would be used to calculate an equivalent profit value. Use "straight line" interpolation.

2. Write a general BASIC program corresponding to your flowchart for exercise 1. The program should be capable of handling 500 sets of profit and sales values. Make up your own data, but for testing use only 20 values in the profit and sales lists. Use -1 for "profit" as the dummy end-of-file indicator.

Lesson 11

LOOPS WITHIN LOOPS

Lesson 11 and 12 may be of interest only to math or science students. Other students may therefore skip them.

Assume you have two arrays, A and V, which look like this:

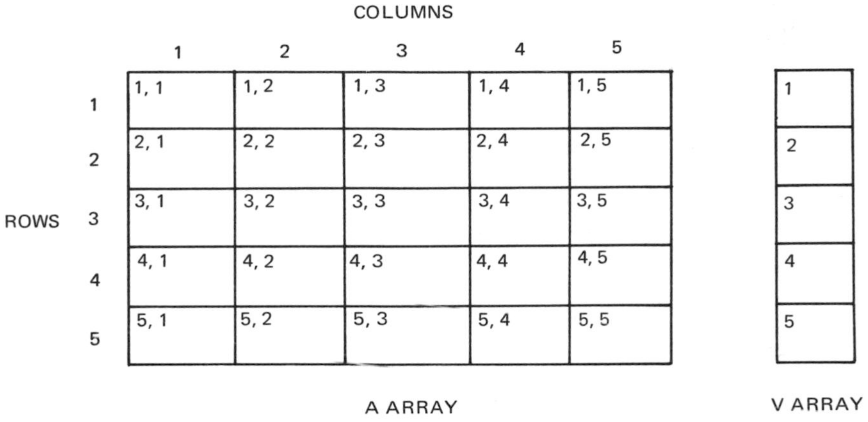

COLUMNS

		1	2	3	4	5
	1	1, 1	1, 2	1, 3	1, 4	1, 5
	2	2, 1	2, 2	2, 3	2, 4	2, 5
ROWS	3	3, 1	3, 2	3, 3	3, 4	3, 5
	4	4, 1	4, 2	4, 3	4, 4	4, 5
	5	5, 1	5, 2	5, 3	5, 4	5, 5

1
2
3
4
5

A ARRAY   V ARRAY

Figure 11-1

You'll notice that every box in both arrays has been identified with subscripts in the upper left-hand corner. That is, box 1,1 is the one in the upper-left hand corner of the array; box 1,5 is the one in the upper right-hand corner; etc. Subscripts give the row first, then the column. Study the figure carefully; you'll need this information for the following problem.

You wish to load both arrays with values so that the resulting arrays look like this. (We haven't shown the subscripts identifying each box. (They're the same as shown in Figure 11-1.)

COLUMNS

	1	2	3	4	5		
1	6	3	4	0	2		3
2	8	5	3	8	3		6
ROWS 3	4	8	9	7	2		5
4	6	0	1	5	3		4
5	5	4	9	2	6		1

A ARRAY   V ARRAY

Figure 11-2

Note that in array A, the box identified as $A_{3,2}$ contains the value, 8 (row 3, column 2); in array V, the box identified as V_3 contains the value, 5. The problem is as follows:

(a) Compute these five sums and print them out:

$$A_{1,1} \times V_1 + A_{1,2} \times V_2 + A_{1,3} \times V_3 + A_{1,4} \times V_4 + A_{1,5} \times V_5$$

$$A_{2,1} \times V_1 + A_{2,2} \times V_2 \text{ etc.}$$

$$A_{3,1} \times V_1 + A_{3,2} \times V_2 \text{ etc.}$$

$$A_{4,1} \times V_1 + A_{4,2} \times V_2 \text{ etc.}$$

$$A_{5,1} \times V_1 + A_{5,2} \times V_2 \text{ etc.}$$

(b) Compute a total of the five sums shown and print it out. Employ a method that uses subscripts and a loop within a loop.

Give this problem a try before you check the flowchart in Figure 11-3.

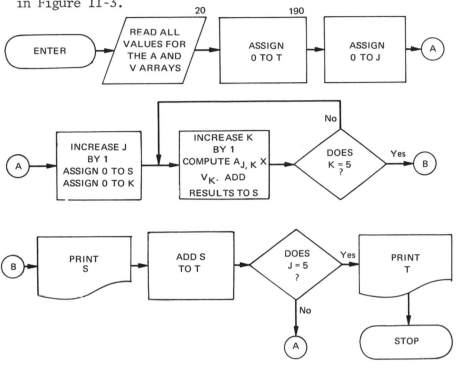

Figure 11-3

The corresponding BASIC program begins like this:

```
 10   DIM A(5,5), V(5)
 20   LET K = 0
 30   LET K = K + 1
 40   LET N = 0
 50   LET N = N + 1
 60   READ A(N,K)
 70   IF N = 5 THEN 90
 80   GO TO 50
 90   IF K = 5 THEN 140
100   GO TO 30
```

(Continued on next page)

```
110   DATA 6, 8, 4, 6, 5, 3, 5, 8, 0, 4, 4, 3, 9
120   DATA 1, 9, 0, 8, 7, 5, 2, 2, 3, 2, 3, 6,
130   DATA 3, 6, 5, 4, 1
140   LET L = 0
150   LET L = L + 1
160   READ V(L)
170   IF L = 5 THEN 190
180   GO TO 150
```

Exercise

1. Complete this program beginning with line 190. (In the
    flowchart S is the individual sum, T is the sum of all
    individual sums.)

Lesson 12

ZEROING IN

In some classes of problems, a program makes a guess
regarding the correct solution to a problem. You then have
the program improve the guess until the solution is found
with the accuracy desired. A simple example is the way that
a computer calculates square root. It uses a form of the
equation

$$G_n = \frac{G_o + \frac{N}{G_o}}{2}$$

where N is the number for which you want the square root,
G_o is the *old guess*, and G_n is the *new guess*.

Given a value of N, the computer establishes an *initial*
value for the old guess, G_o. It can do this in a number of
ways; one of the simplest is to divide N by 2.

The program computes G_n. If it is the same as G_o (within a certain tolerance), then the problem is solved. If not,
G_n is assigned to G_o and another evaluation of the equation
is performed. This procedure continues until G_n and G_o have
the same value within the tolerance required.

Figure 12-1 shows the flowchart.

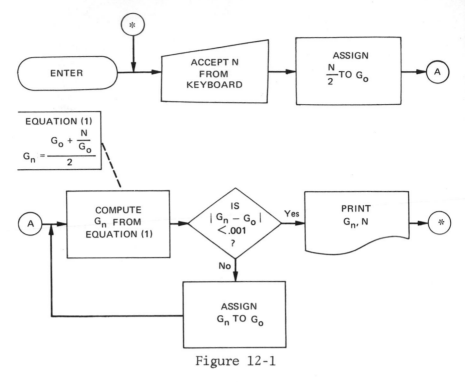

Figure 12-1

Now the program is easy to write:

```
10   INPUT N
20   LET G1 = N/2
30   LET G2 = (G1 + (N/G1))/2
40   IF ABS (G1 - G2) < .001 THEN 70
50   LET G1 = G2
60   GO TO 30
70   PRINT G2, N
80   GO TO 10
```

You can see that G1 is G_o and G2 is G_n. (In BASIC, variable names may be only single letters or single letters followed by single digits.)

Most programming languages have a built-in instruction to compute square roots, so don't use the method just shown.

For example, you'd write

$$P = SQRT \ (N) \ in \ FORTRAN$$
$$LET \ P = SQR \ (N) \ in \ BASIC$$
$$COMPUTE \ P = N \ ** \ .5 \ in \ COBOL$$

etc.

As another example, Figure 12-2 illustrates a well-known ladder problem:

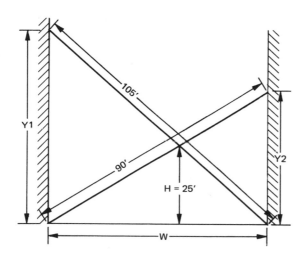

Figure 12-2

Two ladders are leaning against opposite walls in an alley. One is 105 feet long; the other is 90. The height above the ground where they cross is 25 feet. How wide is the alley?

Given this equation:

$$H = \frac{Y_1 \times Y_2}{Y_1 + Y_2}$$

have the program make a guess for W, say 50. Then have it make two other guesses, say 50 - 10 = 40 and 50 + 10 = 60. Then have the program find which guess is best (which guess

gives H closest to 25). Now, have the program make three
more guesses based upon the result of the first evaluation.
Continue this procedure until the answer is found.

Suppose for example, that H is closest to 25 when W is
60. Have the computer select the values 50, 60, and 70 as
the *next three* values of W guesses. Than have the program
find which of *those* values is closest to 25, etc.

Let's begin with an interval of 10 between each W guess.
Have the program *halve* the interval whenever the middle W
gives H closest to 25. Have the program stop searching for
the best W when H is equal to 25 within ± .005. Figure 12-3
shows a flowchart we can use.

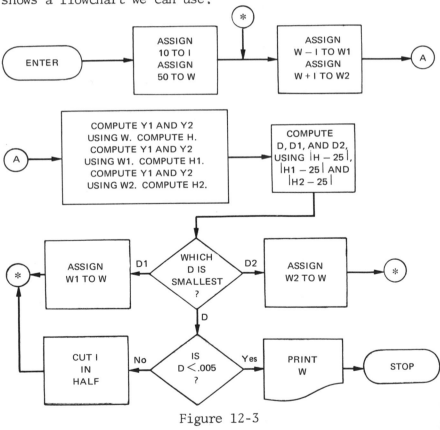

Figure 12-3

Definitions

I   is interval size; initially 10

W   is mid-value of 3 W guesses

W1 is left-most of 3 W guesses

W2 is right-most of 3 W guesses

H
H1   } three calculated H values using
H2   W, W1, and W2, respectively

D
D1   } three absolute value differences from 25
D2   using H, H1, and H2, respectively

The illustration shows Y1 and Y2. They are, of course, computed by using the Pythagorean theorem:

$$c^2 = a^2 + b^2$$

where

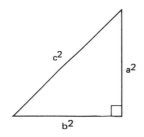

Figure 12-4

Exercises

1. Write the BASIC program to correspond with the flowchart in Figure 12-3.

Lesson 13

EFFICIENT LIST SEARCHING

In a previous lesson, we briefly discussed how to search lists. Since the ability to search lists is very important in programming, we will explore the subject in more detail.

Figure 13-1 shows a simple but <u>inefficient</u> method of list searching.

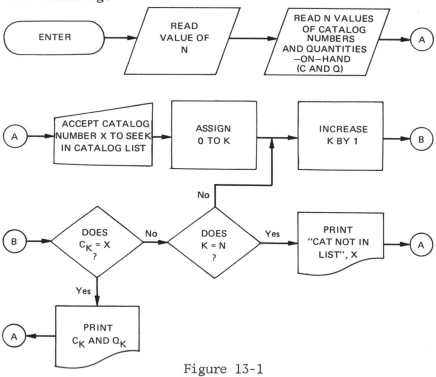

Figure 13-1

Here's the program in BASIC:

```
10    DIM C(500), Q(500)
20    READ N
30    LET K = 0
40    LET K = K + 1
50    READ C(K), Q(K)
60    IF K = N THEN 80
70    GO TO 40
80    INPUT X
90    LET K = 0
100   LET K = K + 1
110   IF C(K) = X THEN 160
120   IF K = N THEN 140
130   GO TO 100
140   PRINT "CAT NOT IN LIST",X
150   GO TO 80
160   PRINT C(K), Q(K)
170   GO TO 80
180   DATA 10, 26, 140, 27, 85, 31, 63, 35,
190   DATA 71, 38, 246, 43, 49, 50, 74, 53
200   DATA 8, 59, 233, 67, 325
```

You've probably recognized that this is a simplified inventory problem. The data gives catalog numbers and quantities. That is, the part having catalog number 26 shows 140 in stock; catalog 27 shows 85 in stock; etc.

Note that the value N shows how many catalog quantity sets are to be used in the problem. In the example, there are only 10 sets used, despite the fact that the DIM statement shows as many as 500 *could* be used. Values of X, catalog numbers to find, are accepted from a teletypewriter.

This is not an especially good way to search a list because the program always begins searching for each new X *at the beginning* of the list. It would be better if the input values of X were in increasing sequence, as the C values already are. Then the program could begin at a point in the C list from where it left off the last time. Only a slight change is needed in the flowchart (see Figure 13-2).

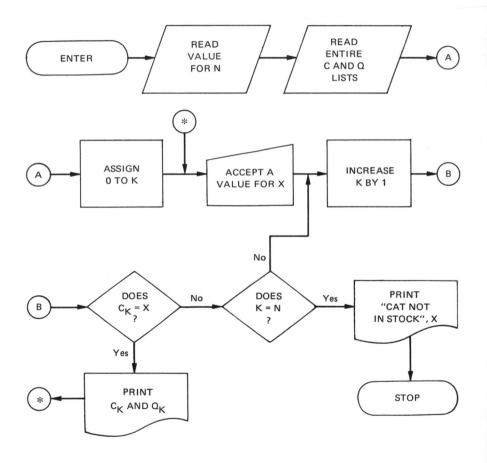

Figure 13-2

A particularly effective way to search a list when input values are in *random* order is the "binary search." The idea is to try to find X in the midpoint of the list, then, if it is not found, to divide the remaining parts of the list in half and to keep doing this until X is found. Suppose for example, you have a list of 1000 values. You need to find value 320. Assume that location 500 of the list holds the value 676. You now know that the value desired, 320, must lie between locations 1 and 449, inclusive.

See if you can write a flowchart to show how a binary search is conducted. For simplicity, let's include only 20 sets of values in the list. The sets have this form:

$$C_1, Q_1; C_2, Q_2; \ldots; C_{20}, Q_{20}$$

where C is a list of catalog numbers and Q is a corresponding list of quantities-on-hand. Both lists are in numerically increasing sequence. Have the program write a message if the required catalog number is not in list C. The flowchart is shown in Figure 13-3.

Exercise

1. Write the BASIC program for the binary search problem described above.

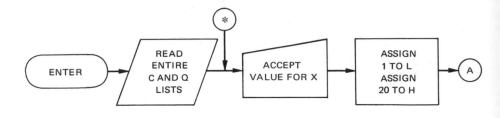

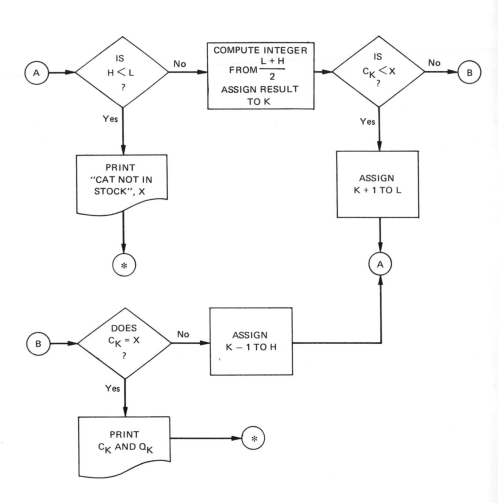

Figure 13-3

Lesson 14

FILE MAINTENANCE

In business data processing, it is often necessary to up-
date a file. In practice, this process can be extremely complex, but in this lesson we'll give a simplified version. We
need a new symbol to represent file. Inasmuch as files are
often on magnetic tape, we'll introduce the *magnetic tape*
symbol.

When files are read, complete units called *records* are
made available to a program. A record is a set of related
data items. For example, a single record will give one person's social security number, name, year-to-date earnings,
job description, pay rate, etc.

Business programs are usually coded in COBOL or PL/1,
but other languages can also be used. Even BASIC has a set
of file-handling instructions. The system does not permit
the actual accessing of magnetic tapes, but tapes can be sim-
ulated. Here's the BASIC program that corresponds to the
flowchart shown in Figure 14-1.

```
10    FILES A; B
20    SCRATCH #2
30    IF END #1 THEN 70
40    READ #1, T, U, V, W, X, Y, Z
50    WRITE #2, T; U; V; W; X; Y; Z
60    GO TO 30
70    STOP
```

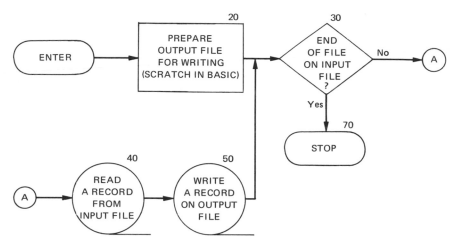

Figure 14-1

File A is a saved master file. It is referred to as #1
in the program. File B is a *blank* output file. It is #2.
Observe that when a READ is written, all the named data items
in the instruction are separated by *commas*. When a WRITE is
written, all the named data items in the instruction are
separated by *semicolons*. Note also that, as a safety mea-
sure, you are required to SCRATCH a file before you can have
the program write upon it. Finally, observe in line 30 how
an "end-of-file" condition is tested.

In the preceding example, not much is being done. A file
on magnetic tape is being read, record-by-record, and is

being copied onto another reel of tape. Now let's now discuss a more meaningful problem. Assume you have a master file called MASTER. It has only two data items per record: a part number and a quantity. Assume also that you have a transaction file called TRANS. It has three data items per record: a part number, a transaction code, and a quantity. The transaction codes are:

1 meaning "this is a record to add"

2 meaning "this is a record to delete"

3 meaning "this record shows a disbursement from stock"

Where the transaction code is 1, the quantity shown in the record is the quantity-on-hand for the new record; where the transaction code is 2, the quantity shown in the record is zero (the item is to be dropped from stock; therefore, quantity information is irrelevant); where the transaction code is 3, the quantity shown in the record is the disbursement quantity (this quantity must be subtracted from the quantity-on-hand shown in the master file).

A third file used in the problem is NMF, meaning NEW MASTER FILE. When the program begins, this file is blank.

Records are filed in the master and transaction files in increasing sequence by parts numbers. In processing, records are read from both the master and transaction files. The program looks for matches in part numbers. If a match is detected, the transaction file is checked to see whether the transaction code is 2 (meaning delete) or 3 (meaning disbursement). If the transaction code 2, the corresponding master record is dropped; that is, it is *not* copied onto NMF. If the file code is 3, the quantity shown in the transaction record is subtracted from the quantity on hand shown in the

-85-

master record. Then the updated master record is *copied* onto NMF. If the transaction code is not 2 or 3, an error has
occurred and the program must write out an error message.

When part number keys do not match, the program must de-
cide whether the reason is that there existed no transaction
applying to a master record or whether there is a new record
to be added to NMF. If the former is true, the master record must be copied without change onto NMF. If the latter
is true, the program must copy the entire transaction record
(except for the transaction code) onto NMF *after* having
checked the transaction code to make sure it is 1 (meaning,
add this record to the new master file). If that transaction
code is not 1, an error message must be written out. You
should try to work out a flowchart for this problem before
referring to the flowchart shown in Figure 14-2.

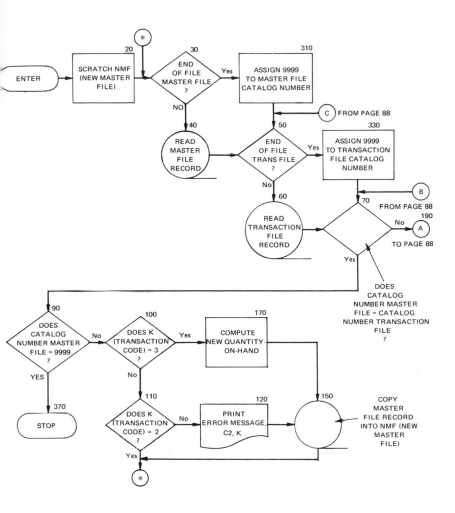

Figure 14-2

(Continued on next page)

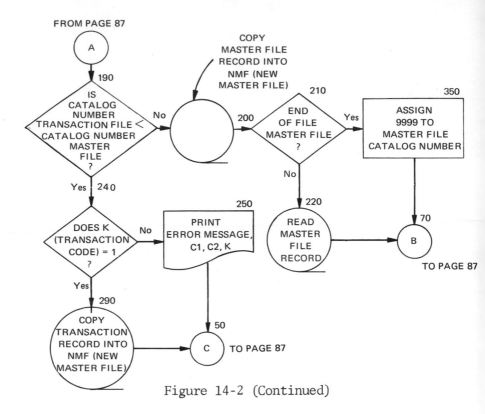

Figure 14-2 (Continued)

The corresponding BASIC program is shown as follows:

```
10    FILES MASTER; TRANS; NMF
20    SCRATCH #3
30    IF END #1 THEN 310
40    READ #1, C1, Q1
50    IF END #2 THEN 330
60    READ #2, C2, Q2, K
70    IF C1 = C2 THEN 90
80    GO TO 190
90    IF C1 = 9999 THEN 370
100   IF K = 3 THEN 170
110   IF K = 2 THEN 30
```

(Continued on next page)

```
120   PRINT "ERR. CAT NUMBS ARE EQUAL BUT";
130   PRINT " TRANS CODE NOT DISB OR DEL",
140   PRINT C2; K
150   WRITE #3, C1; Q1
160   GO TO 30
170   LET Q1 = Q1 - Q2
180   GO TO 150
190   IF C2 < C1 THEN 240
200   WRITE #3, C1; Q1
210   IF END #1 THEN 350
220   READ #1, C1, Q1
230   GO TO 70
240   IF K = 1 THEN 290
250   PRINT "ERR TRANS CAT NUM < MAST CAT";
260   PRINT " NUM BUT TRANS CODE NOT 1",
270   PRINT C1; C2; K
280   GO TO 50
290   WRITE #3, C2; Q2
300   GO TO 50
310   LET C1 = 9999
320   GO TO 50
330   LET C2 = 9999
340   GO TO 70
350   LET C1 = 9999
360   GO TO 70
370   STOP
```

In the program, C1 is the part number key appearing in master file records; C2 is the part number key appearing in transaction file records. By the same token, Q1 and Q2 are quantities appearing in master and transaction file records, respectively. Transaction file records also show the data item, K, which represents the transaction code.

At line 170, Q1 is computed. This is the new quantity-on-hand to write in the new master file record. This value is computed only when record keys match and when the transaction code is 3.

Observe that when either the master file or the transaction file runs out of records, the catalog number being held in memory for either of the files is set to 9999. This number is a *dummy* catalog number larger than any actual catalog number. The program stops when the catalog number of the master file in memory equals the catalog number in memory of the transaction file *and* both catalog numbers equal 9999.

In the program, the complete nomenclature is:

C1 catalog number, master file

C2 catalog number, transaction file

Q1 quantity-on-hand, master file

Q2 quantity-on-hand, transaction file

K   transaction code

#1 name of master file (MASTER)

#2 name of transaction file (TRANS)

#3 name of new master file (NMF)

Answers

Lesson 1

1.

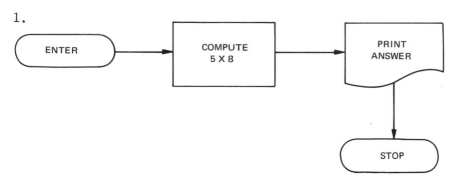

2. The arrow between COMPUTE and PRINT points the wrong way.

3. The arrow between COMPUTE and PRINT, and the arrow between PRINT and STOP point the wrong way.

4.

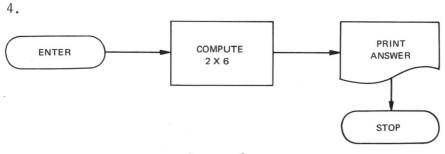

Lesson 2

1.

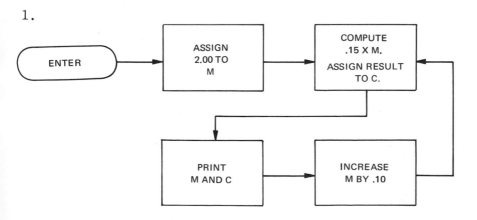

2.

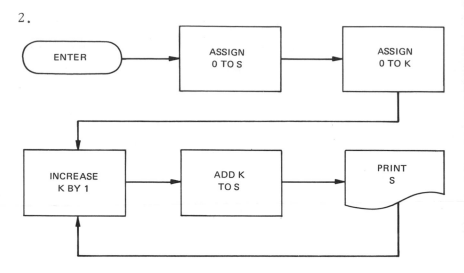

3. The arrow from "PRINT P" should go to INCREASE THE VALUE OF P BY 1. The way it is now, you would receive a printout of 1's only.

4.

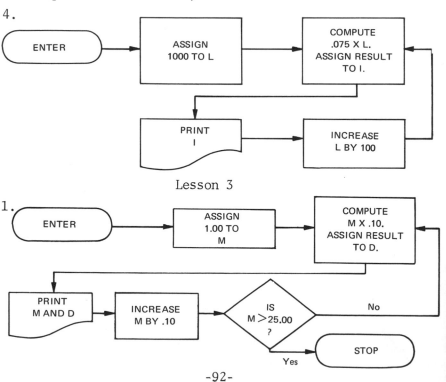

Lesson 3

1.

2.

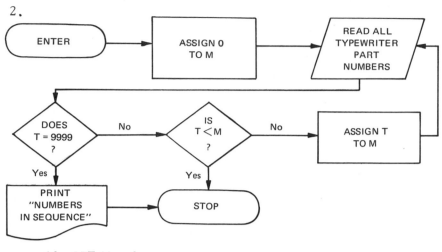

```
10   LET M = 0
20   READ T
30   IF T = 9999 THEN 70
40   IF T < M THEN 80
50   LET M = T
60   GO TO 20
70   PRINT "NUMBERS IN SEQUENCE"
80   STOP
90   DATA 3014, 3029, 3083, 4017, 4044, 9999
```

3.

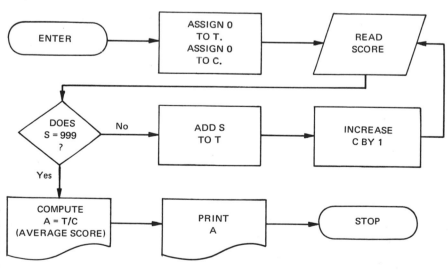

3. (Continued)

```
10   LET T = 0
20   LET C = 0
30   READ S
40   IF S = 999 THEN 80
50   LET T = T + S
60   LET C = C + 1
70   GO TO 30
80   LET A = T/C
90   PRINT A
100  STOP
110  DATA 55, 60, 65, 70, 75, 80, 85, 90, 95, 100, 999
```

4.

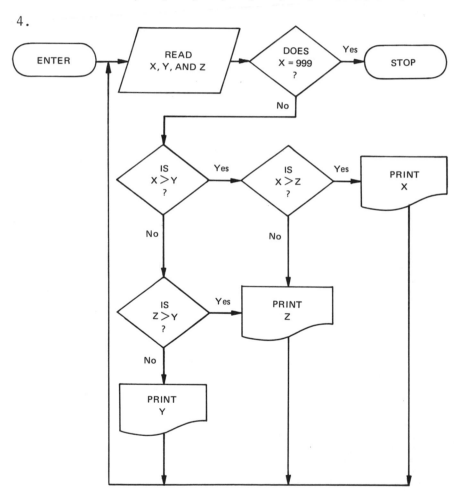

4. (Continued)

```
10    READ X, Y, Z
20    IF X = 999 THEN 120
30    IF X > Y THEN 70
40    IF Z > Y THEN 80
50    PRINT Y
60    GO TO 10
70    IF X > Z THEN 100
80    PRINT Z
90    GO TO 10
100   PRINT X
110   GO TO 10
120   STOP
130   DATA 2, 4, 6, 8, 7, 9, 8, 4, 3, 5, 8, 3, 999, 0, 0
```

Lesson 4

1.

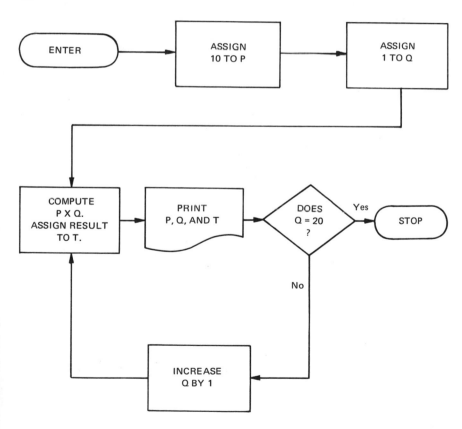

2.  ```
    10   LET P = 10
    20   LET Q = 1
    30   LET T = P * Q
    40   PRINT P, Q, T
    50   IF Q = 20 THEN 80
    60   LET Q = Q + 1
    70   GO TO 30
    80   STOP
    ```

3.

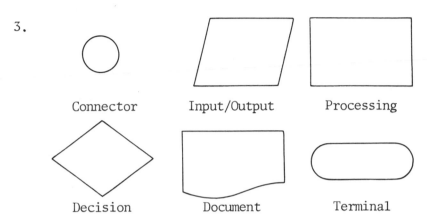

Connector   Input/Output   Processing

Decision   Document   Terminal

Lesson 5

1.

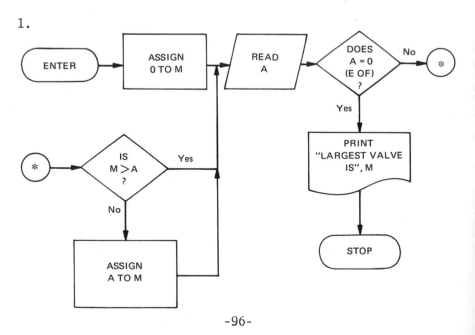
    ```

2. Yes.

Lesson 6

1.

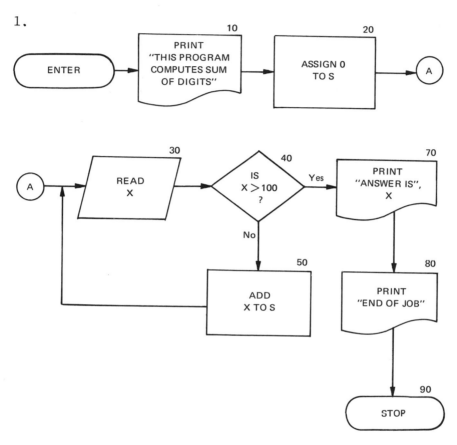

2. a. Input data read at line 30.
    b. Data processed at line 50.
    c. Answers written out at line 70.
    d. End-of-file tested at line 40.
    e. Initialization at lines 10 and 20.

3.

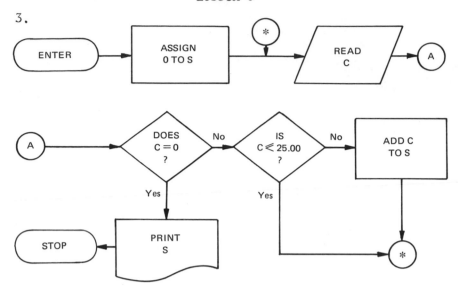

4. 10   LET S = 0
    20   READ C
    30   IF C = 0 THEN 70
    40   IF C < = 25.00 THEN 20
    50   LET S = S + C
    60   GO TO 20
    70   PRINT S
    80   STOP
    90   DATA 25.50, 26.00, 23.00, 20.00, 40.00, 0

Lesson 7

1.

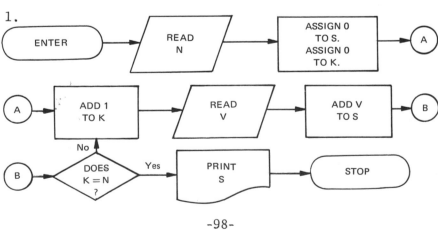

2.

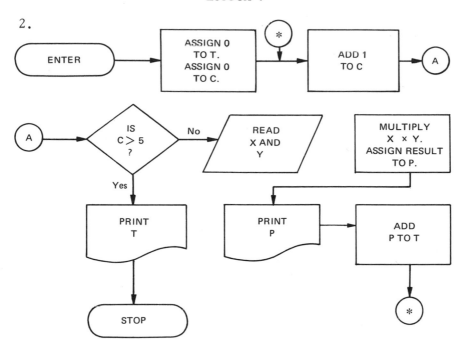

3.

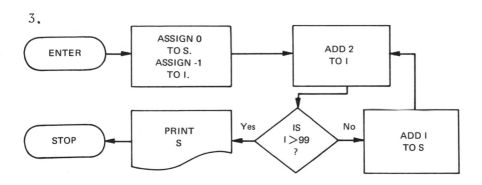

4.
```
10   LET S = 0
20   LET I = -1
30   LET I = I + 2
40   IF I > 99 THEN 70
50   LET S = S + I
60   GO TO 30
70   PRINT S
80   STOP
```

5.

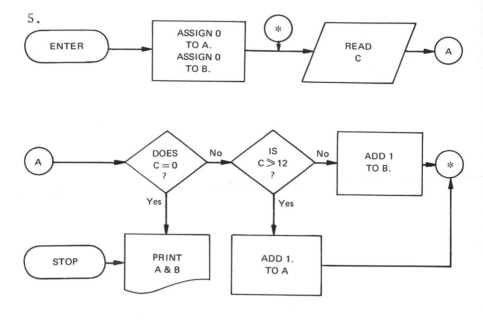

```
10    LET A = 0
20    LET B = 0
30    READ C
40    IF C = 0 THEN 100
50    IF C > 12 THEN 80
60    LET B = B + 1
70    GO TO 30
80    LET A = A + 1
90    GO TO 30
100   PRINT A, B
110   DATA 2, 13, 4, 6, 24, 45, 1, 7, 15, 16, 6, 0
```

NOTE: A contains the number
of ages that are 12
or above; B contains
the number of ages
below 12.

6.

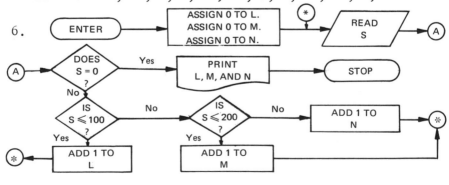

6(Continued)

```
10   LET L = 0
20   LET M = 0
30   LET N = 0
40   READ S
50   IF S = 0 THEN 140
60   IF S < = 100 THEN 100
70   IF S < = 200 THEN 120
80   LET N = N + 1
90   GO TO 40
100  LET L = L + 1
110  GO TO 40
120  LET M = M + 1
130  GO TO 40
140  PRINT L, M, N
150  STOP
160  DATA 50, 110, 210, 150, 75, 250, 600, 0
```

NOTE:
L contains the number of sales which were equal to $100 or less; M contains the number of sales which were greater than $100 but $200 or less; M contains the number of sales which were greater than $200.

Lesson 8

1.
```
10   LET X = 0
20   LET T = 0
30   LET S = (3.1416/2)/100
40   LET Z = 3.1416/2
50   LET X = X + S
60   IF X > Z THEN 100
70   LET A = SIN(X) * S
80   LET T = T + A
90   GO TO 50
100  PRINT T
110  STOP
```

NOTE:
Program runs faster because $\pi/100$ and $\frac{\pi}{2}$ are computed only once.

2.
```
10   LET X = 0
20   LET T = 0
30   LET X = X + (3.1416/2)/100
40   IF X > 3.1416/2 THEN 80
50   LET T = T + (SIN(X) * ((3.1416/2)/100))
70   GO TO 30
80   PRINT T
90   STOP
```

3.

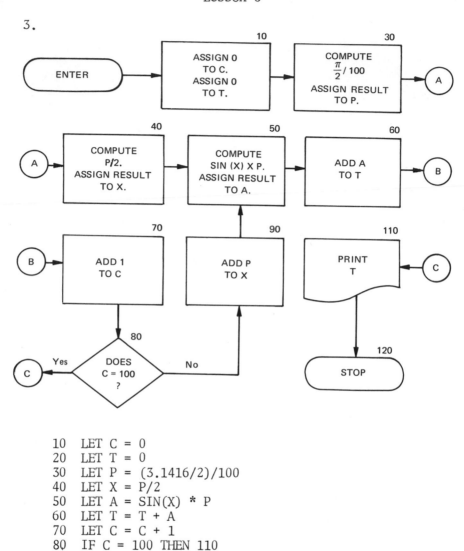

```
10   LET C = 0
20   LET T = 0
30   LET P = (3.1416/2)/100
40   LET X = P/2
50   LET A = SIN(X) * P
60   LET T = T + A
70   LET C = C + 1
80   IF C = 100 THEN 110
90   LET X = X + P
100  GO TO 50
110  PRINT T
120  STOP
```

4.

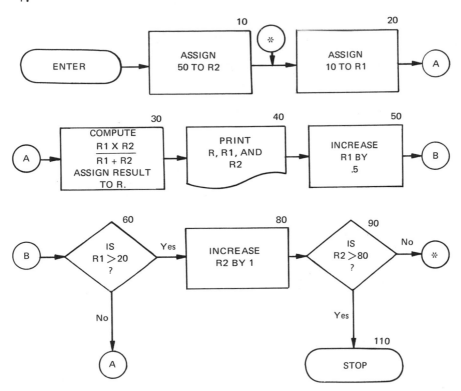

5. 10   LET R2 = 50
    20   LET R1 = 10
    30   LET R = (R1 * R2)/(R1 + R2)
    40   PRINT R; R1; R2
    50   LET R1 = R1 + .5
    60   IF R1 > 20 THEN 80
    70   GO TO 30
    80   LET R2 = R2 + 1
    90   IF R2 > 80 THEN 110
   100   GO TO 20
   110   STOP

Lesson 9

1.

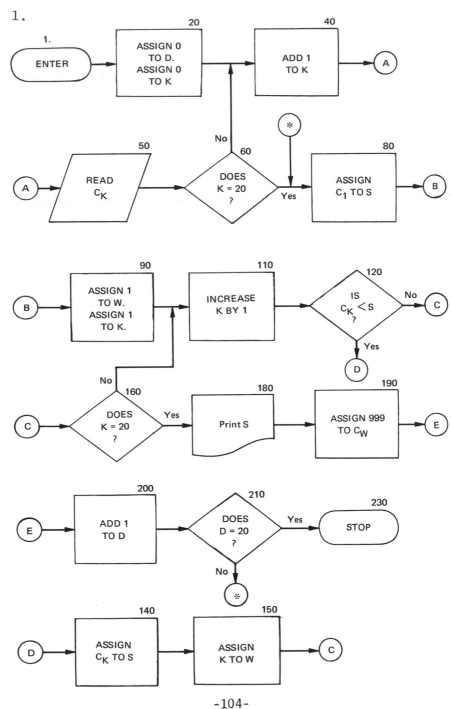

1.

ENTER → ASSIGN 0 TO D. ASSIGN 0 TO K [20] → ADD 1 TO K [40] → (A)

(A) → READ c_K [50] → DOES K = 20 ? [60] → No (up to 20) / Yes → * → ASSIGN c_1 TO S [80] → (B)

(B) → ASSIGN 1 TO W. ASSIGN 1 TO K. [90] → INCREASE K BY 1 [110] → IS $c_K < S$? [120] → No → (C) / Yes → (D)

(C) → DOES K = 20 ? [160] → No (up to 110) / Yes → Print S [180] → ASSIGN 999 TO c_W [190] → (E)

(E) → ADD 1 TO D [200] → DOES D = 20 ? [210] → Yes → STOP [230] / No → *

(D) → ASSIGN c_K TO S [140] → ASSIGN K TO W [150] → (C)

-104-

Lesson 9

```
2.   10   DIM C(20)
     20   LET D = 0
     30   LET K = 0
     40   LET K = K + 1
     50   READ C(K)
     60   IF K = 20 THEN 80
     70   GO TO 40
     80   LET S = C(1)
     90   LET W = 1
    100   LET K = 1
    110   LET K = K + 1
    120   IF C(K) < S THEN 140
    130   GO TO 160
    140   LET S = C(K)
    150   LET W = K
    160   IF K = 20 THEN 180
    170   GO TO 110
    180   PRINT S
    190   LET C(W) = 999
    200   LET D = D + 1
    210   IF D = 20 THEN 230
    220   GO TO 80
    230   STOP
    240   DATA 3, 9, 4, 17, 8, 35, 12, 11, 42, 5, 6
    250   DATA 7, 1, 13, 90, 10, 22, 2, 14, 44
```

NOTE:
S means smallest value;
W means *which* value in
the list is smallest.
List is searched 20 times;
D counts number of times.

Lesson 10

1.

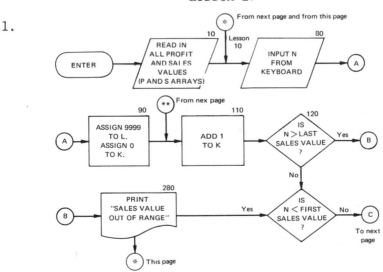

1. (Continued)

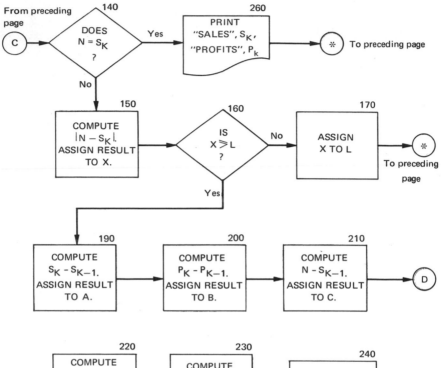

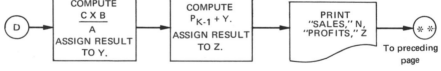

2.
```
10    DIM P(500), S(500)
20    LET C = 0
30    LET C = C + 1
40    READ P(C), S(C)
50    IF S(C) = -1 THEN 70
60    GO TO 30
70    LET C = C - 1
80    INPUT N
90    LET L = 9999
100   LET K = 0
110   LET K = K + 1
120   IF N > S(C) THEN 280
130   IF N < S(1) THEN 280
```

(Continued on Next Page)

2. (Continued)

```
140   IF S(K) = N THEN 260
150   LET X = ABS(N - S(K))
160   IF X > = L THEN 190
170   LET L = X
180   GO TO 110
190   LET A = S(K) - S(K - 1)
200   LET B = P(K) - P(K - 1)
210   LET C = N - S(K - 1)
220   LET Y = (C * B)/A
230   LET Z = P(K - 1) + Y
240   PRINT "SALES"; N, "PROFITS"; Z
250   GO TO 80
260   PRINT "SALES"; S(K), "PROFITS"; P(K)
270   GO TO 80
280   PRINT "SALES VALUE OUT OF RANGE"
290   GO TO 80
300   DATA 24,124,32.5,130,36,136,38,142
310   DATA 40,148,45,150,50.5,170,58,178
320   DATA 62,183,66,190
330   DATA 70,193,73,198,80,200,85,205,88,209,92,212
340   DATA 97,216,100,200,105,215,111,225,0,-1
```

Lesson 11

```
1.    190   LET T = 0
      200   LET J = 0
      210   LET J = J + 1
      220   LET S = 0
      230   LET K = 0
      240   LET K = K + 1
      250   LET S = S + (A(J,K) * V(K))
      260   IF K = 5 THEN 280
      270   GO TO 240
      280   PRINT S
      290   LET T = T + S
      300   IF J = 5 THEN 320
      310   GO TO 210
      320   PRINT T
      330   STOP
```

```
1.   10    LET I = 10
     20    LET W = 50
     30    LET W1 = W - I
     40    LET W2 = W + I
     50    LET Y1 = SQR (105↑2 - W↑2)
     60    LET Y2 = SQR (90↑2 - W↑2)
     70    LET H = (Y1 * Y2)/(Y1 + Y2)
     80    LET Y1 = SQR (105↑2 - W1↑2)
     90    LET Y2 = SQR (90↑2 - W1↑2)
     100   LET H1 = (Y1 * Y2)/(Y1 + Y2)
     110   LET Y1 = SQR(105↑2 - W2↑2)
     120   LET Y2 = SQR(90↑2 - W2↑2)
     130   LET H2 = (Y1 * Y2)/(Y1 + Y2)
     140   LET D = ABS (H - 25)
     150   LET D1 = ABS(H1 - 25)
     160   LET D2 = ABS(H2 - 25)
     170   IF D<D1 THEN 210
     180   IF D1<D2 THEN 230
     190   LET W = W2
     200   GO TO 30
     210   IF D<D2 THEN 250
     220   GO TO 190
     230   LET W = W1
     240   GO TO 30
     250   IF D<.005 THEN 280
     260   LET I = I/2
     270   GO TO 30
     280   PRINT W
     290   STOP

2.   10    DIM C(20), Q(20)
     20    LET K = 0
     30    LET K = K + 1
     40    READ C(K), Q(K)
     50    IF K = 20 THEN 70
     60    GO TO 30
     70    INPUT X
     80    LET L = 1
     90    LET H = 20
     100   IF H<L THEN 200
     110   LET K = INT((L + H)/2)
     120   IF C(K)<X THEN 160
     130   IF C(K) = X THEN 180
     140   LET H = K - 1
     150   GO TO 100
```

(Continued on Next Page)

2. (Continued)

```
160   LET L = K + 1
170   GO TO 100
180   PRINT C(K), Q(K)
190   GO TO 70
200   PRINT "CAT NOT IN STOCK",X
210   GO TO 70
220   DATA 1,10,2,20,3,30,4,40,5,50,6,60,7,70
230   DATA 8,80,9,90,10,100,11,110,12,120,13,130
240   DATA 14,140,15,150,16,160,17,170,18,180
250   DATA 19,190,20,200
```

Index